Information Circular 9475

Ergonomic Assessment of Musculoskeletal Risk Factors at Four Mine Sites: Underground Coal, Surface Copper, Surface Phosphate, and Underground Limestone

By William J. Wiehagen and Fred C. Turin

U.S. DEPARTMENT OF HEALTH AND HUMAN SERVICES
Centers for Disease Control and Prevention
National Institute for Occupational Safety and Health
Pittsburgh Research Laboratory
Pittsburgh, PA

August 2004

ORDERING INFORMATION

Copies of National Institute for Occupational Safety and Health (NIOSH)
documents and information
about occupational safety and health are available from

NIOSH–Publications Dissemination
4676 Columbia Parkway
Cincinnati, OH 45226–1998

FAX: 513–533–8573
Telephone: 1–800–35–NIOSH
(1–800–356–4674)
E-mail: pubstaff@cdc.gov
Web site: www.cdc.gov/niosh

This document is the public domain and may be freely copied or reprinted.

Disclaimer: Mention of any company or product does not constitute endorsement by NIOSH.

DHHS (NIOSH) Publication No. 2004–159

CONTENTS

Page

Abstract	1
Acknowledgments	2
Glossary	2
Background	3
Risk factors	4
Methods and tools	4
Site selection	6
Baseline data	6
Incident data	6
Nordic questionnaire	6
Supervisor interview guide	6
Management concerns	6
Job selection	6
Task analysis	7
Body part discomfort interview guide	7
Brainstorming	7
Results	7
Selecting work activities and tasks	7
MSD risk factors for selected work tasks	7
Summary	11
References	12
Appendix A.—Ergonomic interventions for the mining industry: an overview of planned approach	13
Appendix B.—Nordic questionnaire	15
Appendix C.—Front-line supervisor interview guide	17
Appendix D.—Body part discomfort interview guide	20
Appendix E.—Sample decision matrix	22
Appendix F.—Underground coal mine	23
Appendix G.—Surface copper mine	26
Appendix H.—Surface phosphate mine	30
Appendix I.—Underground limestone mine	34

ILLUSTRATIONS

1.	Overall study strategy	5

TABLES

1.	Mine descriptions, tools used to select tasks, and tasks selected across the four sites	8
2.	Risk factors by observed tasks: underground coal mine	9
3.	Risk factors by observed tasks: surface copper mine	10
4.	Risk factors by observed tasks: surface phosphate mine	10
5.	Risk factors by observed tasks: underground limestone mine	11
E–1.	Sample decision matrix	22
F–1.	Underground coal: number of incidents and employees by job title and incident rate, 1996–1998	23
F–2.	Underground coal: incidents by occupational class and incident type, 1996–1998	24
F–3.	Underground coal: roof bolter operator – installing bolts and manual materials handling	24
F–4.	Underground coal: continuous miner operator – cutting coal	25
F–5.	Underground coal: shuttle car operator – transporting coal	25
F–6.	Underground coal: motormen – unloading supplies	25
G–1.	Surface copper: number of incidents and employees by work area and incident rate, 1996–1998	26
G–2.	Surface copper: incidents by work group and incident type, 1996–1998	27
G–3.	Surface copper: ratings of body part discomfort by work area	27

TABLES–Continued

Page

G–4.	Surface copper: decision matrix for selecting tasks	28
G–5.	Surface copper: smelter operator – converter turnaround	28
G–6.	Surface copper: smelter operator – 4B–5 gun rebuild	28
G–7.	Surface copper: refinery operator – cell checking	29
G–8.	Surface copper: tank house operator – bar pulling	29
H–1.	Surface phosphate: number of incidents and employees by job title and incident rate, 1996–1998	30
H–2.	Surface phosphate: incidents by job title and incident type, 1996–1998	31
H–3.	Surface phosphate: ratings of body part discomfort by work area	31
H–4.	Surface phosphate: physically demanding jobs identified by supervisors	31
H–5.	Surface phosphate: float crew – wrenching pipes	32
H–6.	Surface phosphate: hydraulic pit station operator – operating hydraulic pit station	32
H–7.	Surface phosphate: dozer operation – moving material	33
H–8.	Surface phosphate: reagent operator – emptying reagent cars	33
I–1.	Underground limestone: number of incidents and employees by work area, job title, and incident rate, January 1997–May 2000	34
I–2.	Underground limestone: incidents by work group and incident type, January 1997–May 2000	35
I–3.	Underground limestone: number of reported body parts with discomfort by work group	35
I–4.	Underground limestone: Nordic questionnaire responses by work group	35
I–5.	Underground limestone: physically demanding jobs identified by supervisors	36
I–6.	Underground limestone: decision matrix for selecting tasks	36
I–7.	Underground limestone: quality control technician – gathering and testing samples	37
I–8.	Underground limestone: blasting crew – handling ANFO and dynamite	37
I–9.	Underground limestone: blasting crew – hand scaling and clearing bottom holes	38
I–10.	Underground limestone: scaling machine operator – mechanical scaling of face and back	38

UNIT OF MEASURE ABBREVIATIONS USED IN THIS REPORT

ft	foot		°	degree
in	inch		°F	degree Fahrenheit
lb	pound		%	percent
min	minute			

ERGONOMIC ASSESSMENT OF MUSCULOSKELETAL RISK FACTORS AT FOUR MINE SITES: UNDERGROUND COAL, SURFACE COPPER, SURFACE PHOSPHATE, AND UNDERGROUND LIMESTONE

By William J. Wiehagen[1] and Fred C. Turin[1]

ABSTRACT

This study examined musculoskeletal injury risk at four mining sites: underground coal, underground limestone, surface copper, and surface phosphate. Each site offered opportunities to investigate musculoskeletal disorder (MSD) injury risks and how those risks might be identified and categorized. The National Institute for Occupational Safety and Health (NIOSH) worked with these sites to (1) identify work activities that showed evidence of MSD injury risk, (2) examine physical risk factors that can lead to MSDs for a handful of work tasks at each site, and (3) develop a set of ideas for problem-solving to help reduce risk factors for examined work tasks.

For each site, NIOSH implemented a plan that was refined over the time period of this study. The plan consisted of four steps. The first step was to use mine injury records, a musculoskeletal discomfort questionnaire, front-line supervisor interviews, and a list of management concerns to identify work groups and work activities that have significant evidence of MSD risk factors. The second step was to select work tasks for evaluation. The third step was to interview those who do the work and make observations to characterize the MSD risk factors and musculoskeletal symptoms that exist for target tasks. The final step was to conduct brainstorming sessions with workers who perform the work or have a stake in the production task. The brainstorming sessions served to identify general strategies (ideas) for reducing MSD risk factor exposures.

A final report of findings was presented to mine management and workforce representatives at each site. The risk factors and ideas for improvement identified for each site were specific to the target tasks examined. These target tasks were diverse, but there were some key similarities. For instance, jobs were found at each site that required a significant amount of manual work involving the upper extremities and low back. Handling heavy and awkward objects, forceful arm and shoulder exertions, and working in awkward postures were common for a variety of jobs across the four sites.

---

[1]Industrial engineer, Pittsburgh Research Laboratory, National Institute for Occupational Safety and Health, Pittsburgh, PA.

ACKNOWLEDGMENTS

This report is based on the work of four NIOSH teams. The efforts involved establishing a plan, visiting mine sites to gather site-wide and task-specific information, analyzing and synthesizing information (via observations, interviews, and physical measurements), and preparing site-specific feedback reports. Contributors to the mine site teams were:

Underground coal.—Lisa J. Steiner, Eric R. Bauer, Ph.D., Kim M. Cornelius,[2] Fred C. Turin, Sean Gallagher, Ph.D.;

Surface copper.—Sean Gallagher, Ph.D., Fred C. Turin, William J. Wiehagen, Albert H. Cook;

Surface phosphate.—Eric R. Bauer, Ph.D., Lisa J. Steiner, Lynn L. Rethi,[2] Albert H. Cook, Kim M. Cornelius, Fred C. Turin; and

Underground limestone.—Fred C. Turin, Kim M. Cornelius, William J. Wiehagen, Albert H. Cook.

GLOSSARY

Musculoskeletal disorder (MSD).—A condition or disorder that involves the muscles, nerves, tendons, ligaments, joints, cartilage, or spinal discs. These disorders are not typically the result of a distinctive, singular event, but are more gradual in their development. Thus, MSDs are cumulative-type injuries.

Acute injury.—A singular, traumatic event resulting in a disruption of tissues, resulting in pain [Kumar 2001].

MSD risk factors.—Actions or conditions that increase the likelihood of injury to the musculoskeletal system. Risk factors have components of duration, frequency, and level of exposure. Exposure to MSD risk factors leads to discomfort and pain, which over time can lead to more serious disorders of the musculoskeletal system.

Ergonomics.—A discipline or science of fitting workplace conditions and job demands to the capabilities of the worker. Many consider ergonomics a multidisciplinary field of applied science where knowledge about human capabilities, skills, limitations, and needs is taken into account when examining the interactions among people, technology, and the work environment [Westgaard and Winkel 1997; Cohen et al. 1997].

---

[2]Now with the National Personal Protective Technology Laboratory, National Institute for Occupational Safety and Health, Pittsburgh, PA.

BACKGROUND

Mining is often characterized as being very difficult with challenging conditions. Whether on the surface or underground, the process of extracting minerals can be characterized as dynamic [Steiner et al. 1999; Scharf et al. 2001]. Researchers suggest that the mining workplace itself is in a state of constant change. Dynamic work processes[3] in particular introduce health and safety risks, and these risks are not always constant. Furthermore, these risks are not always treatable through hierarchal models, suggesting a ranked emphasis on engineering controls, administrative controls, and personal protective equipment [Scharf et al. 2001]. Hazards are part of the mining process and offer challenges to the mining community to research and evaluate strategies to help mitigate the hazards. Realistically, the goal is not one of eliminating hazards, but one of reducing the risk of injury. For this to happen, hazards and their associated risk must be recognized by those who perform, manage, or have a stake in the task.

Musculoskeletal disorders (MSDs) have been identified as a significant and costly problem for the mining industry. Stobbe and Bobick [1986] reported that strain and sprain injuries in 1983 and 1984 accounted for 24.0% and 25.2% of all reported injuries for underground coal mining, respectively, and 19.4% and 20.4% of all injuries for underground metal/nonmetal mining, respectively. The U.S. Bureau of Labor Statistics [1991] reported that mining is among the most hazardous occupations in terms of exposure to ergonomic hazards.[4] An analysis of results from the National Occupational Health Survey of Mining (NOHSM) by Zhuang and Groce [1995] found that the magnitude of musculoskeletal overload potential exposures for coal mining was greater than that for metal/nonmetal mining. In this study, the three most common overload conditions for mine workers were (1) bending forward, bending to the side, hyperextending or twisting the neck or back; (2) unsupported abducted elbows, forearms resting on sharp edges, tossing motions at extremes of range of motion, working with elbows above shoulders; and (3) lifting more than 50 lb, unaided. At least 35% of mine workers were potentially exposed to each of these three overload conditions. In another study, Winn et al. [1996] analyzed NOHSM ergonomic hazard data for 24 commodities associated with the metal/nonmetal mining industry. They determined that potential exposures to ergonomic hazards were most likely for the following body parts: neck and/or back; forearms, arms, and shoulders; and fingers and hands. Overall, they believed that the potential exposure to ergonomic hazards for metal/nonmetal miners was high compared to that for nonmining occupations.

In addition to these common MSD exposures, the workforce is aging in many mining sectors. Fotta and Bockosh [2000] found that older (aged 45+) injured workers have the highest median number of days lost per injury. They examined injury data by type of mining operation and by occupation. The cumulative nature of MSDs suggests that older workers may be more at risk because they have potentially many years of exposure to physically demanding work. In addition, older workers often require a longer period of time to recover from injuries. At the same time, older workers usually have a sound understanding of the work process and can likely make salient suggestions to enhance the process, i.e., make it safer and more efficient. Reducing MSD risk factors will enhance the quality of life of miners, have a positive effect on the productive capacity of the mining plan, and reduce medical costs associated with mine operation.

The NOHSM studies led to a series of discussions in 1998 between Mine Safety and Health Administration (MSHA) Technical Support personnel and NIOSH researchers. MSHA was interested in learning more about the prevalence of MSDs in the mining industry and what might be done to reduce the risk of injury. MSHA expressed the need for exploratory field research to get a better understanding of MSD risk factors at a variety of surface and underground mining sites. These discussions recognized the unique characteristics of mine sites and the wide variety of technology, production processes, and conditions that typify mining. Injuries to the musculoskeletal system (especially back injuries) are common in many mining environments. Identifying and reducing risk factors that can lead to MSDs is a proactive approach to prevent acute and chronic MSDs.

Based on these discussions, NIOSH initiated a project entitled "Ergonomic Interventions in Mining," which began in the summer of 1998. The project consists of two phases. The goal of the first phase was to conduct MSD physical risk factor assessments at four mining sites: underground coal, underground limestone, surface copper, and surface phosphate. NIOSH worked with these sites to (1) identify work tasks that could significantly benefit from additional ergonomic evaluation, (2) examine and categorize physical risk factors that can lead to MSDs for the identified work tasks, and (3) develop a set of ideas for problem-solving to help reduce risk factors for that small set of work tasks examined at each of the four sites.

This report presents the findings from the first phase. Completion was marked by the feedback report being written and presented to the fourth mine site. The study results are presented in a series of tables organized by mining sector and the work task studied. Appendices are provided to offer more detailed information about the specific methods and tools used and additional results from each of the four sites.

---

[3] Examples of dynamic work environments include mining, construction, agriculture, and transportation.

[4] The term "ergonomic hazards" refers to musculoskeletal injury risk due to extended exposure to MSD risk factors.

RISK FACTORS

Risk and risk factors are common concepts used in safety and applied ergonomics literature. Risk includes a component of (1) how likely (i.e., what the probability of) an event is and (2) the seriousness of the consequence (i.e., what the severity is) if something does occur. Risk is often defined in the past tense—looking back and deciding on how many injuries or accidents resulted for a given exposure. At the extremes, injury risk can be viewed as (1) very low probability but extremely high consequence (e.g., multiple fatalities) or (2) higher probability but less severe consequence (e.g., a worker slipping and tripping). Risk is also intuitively relative within and across work settings. Risk implies a probability for injury, and the odds of an injury are a function of the level of risk and worker exposure time. It is possible for workers at a site not to have injuries for a period of time. The absence of injuries does not imply the absence of risk.

In the context of this study, risk factors are defined as actions or conditions that increase the likelihood of injury to the musculoskeletal system. Applied ergonomics literature [Cohen et al. 1997; Washington State Department of Labor 1994] recognizes a small set of common physical risk factors across many occupations and work settings. The relationship between risk factor exposures and the level of musculoskeletal injury risk is not easily defined. Although physical risk factors are important first-line risk factors, there are other plausible factors such as organizational and psychosocial factors that may provoke a disorder or indirectly influence the effect of physical risk factors [Hagberg et al. 1995]. However, there is general agreement that multiple risk factors put a worker at greater risk, even though the increase (via a mathematical probability) in risk cannot be specified. Workers vary in their capacity and ability to adjust to job demands; thus, some may be more affected than others by exposure to the same risk factors [Cohen et al. 1997]. Age also has an effect on variability in worker capacity and ability to adjust to job demands.[5]

Compared to most other work settings, mining offers additional risk complexities due to the dynamic nature of the work. The environment changes daily, and new hazards are introduced as part of the mining process. Work procedures often change to accommodate changes in the work environment. Thus, risk factor exposure is not a constant; it varies depending on the nature of the task, the conditions that surround the task, and the people performing the task.[6]

This study concerns the identification of physical risk factors, within a sample of mining tasks, that can contribute to MSDs/injuries. MSD risk factors[7] that are common to mining include—

- Prolonged, awkward postures
- Forceful exertions (heavy or frequent lifting)
- Forceful gripping
- Highly repetitive motions
- Jolting/jarring
- Vibration exposure (hand and arm)
- Contact stress (pressure points/impact stress)

Risk factor exposure is an early warning of progressively more serious problems—physical signs and symptoms that can lead to serious injury. Long-term exposure to risk factors (e.g., heavy or frequent lifting in awkward postures) will reduce the quality of life.

Every job carries risk. The key issue is relative risk. Organizations and individuals can become better informed to reduce MSD injury risk by (1) being aware of risk factors, (2) becoming skilled in recognizing and categorizing these factors, and (3) examining options to reduce the frequency or duration of exposure to the risk factors. Reducing exposure to risk factors should make the task smoother and more predictable in its outcome. Reducing risk factor exposure should make task performance less variable.

The first step in reducing risk is to identify those work activities that involve significant levels of the seven risk factors, or combinations of those factors, identified above. Exposure to risk factors not only increases the risk of serious injury, but can also routinely inhibit orderly production.

METHODS AND TOOLS

A general protocol (see appendix A) was developed and used by the NIOSH team. It helped to guide the field research effort once a mine had been contacted and expressed interest in participating in the study. A NIOSH project team member served as the site coordinator, and a site evaluation team was formed for each mine. Each of the four NIOSH teams, using the general protocol as a guide, developed a site-specific approach tailored to the conditions and constraints encountered at each site.

This section provides a general overview of the study methods and tools. Because of the exploratory nature of this study, the data collection activities at each site were slightly and necessarily different. However, figure 1 represents what the NIOSH team considers to be one useful strategy for site-wide assessment of MSD injury risk. This assessment leads to (1) selecting work activities that would be good candidates for further study, (2) working with mine site personnel to better

---

[5]The effect of "age" was not examined under this study. However, age is a consideration, as there are a lot of normative changes that can and do occur with aging. For a complete discussion of age and its potential effect on work performance capacities and abilities, see Hagglund [1998].

[6]This is a key reason why training and education in recognizing MSD risk factors is very important.

[7]Duration of tasks and environmental conditions also contribute to the likelihood of injury.

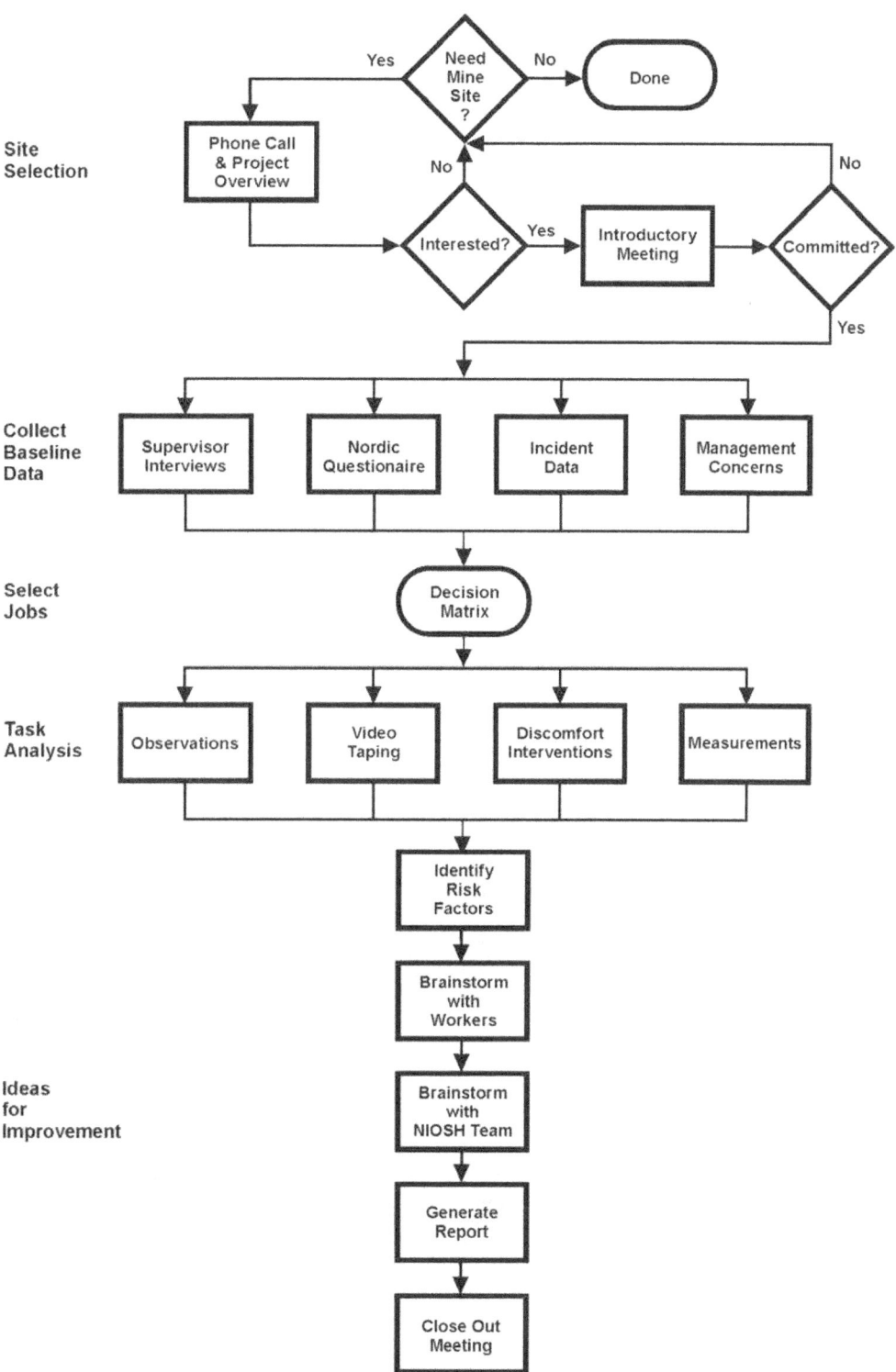

Figure 1.—Overall study strategy.

define the MSD injury risk for specific tasks, and (3) exploring ideas for improving the work process. Improving the work process implies reducing injury risk, i.e., helping to make some tasks less physically demanding or stressful.

SITE SELECTION

From personal contacts, NIOSH team members identified a list of potential sites.[8] A member of the NIOSH team communicated with a mine contact to provide an overview of the project goals and requirements. If interest was expressed, an overview description was sent to the mine contact to brief the mine's management team (see appendix A). If the logistics of this work appeared to be feasible, an introductory meeting was held with management and workforce representatives, along with a representative from MSHA's Technical Support group. Following this meeting, if a mine agreed to cooperate, the mine site contact person and the NIOSH team member planned the details to facilitate the study.

BASELINE DATA

To establish a baseline at each site, the NIOSH team used a combination of data collection tools: (1) incident data, (2) Nordic Questionnaires, (3) first-line supervisor interviews, and (4) management concerns. The overall intent of gathering baseline data was to form a site-wide picture of relative MSD injury risk by work activity. To accomplish this, the NIOSH team used both qualitative and quantitative data collection tools. This met two objectives: (1) it helped the NIOSH team learn more about the mining technology, process, and conditions present at each site, and (2) it allowed the team to collaborate with workers, front-line supervisors, and management to develop an objective and logical structure for selecting work tasks for in-depth study.

Incident Data

At least 3 years of incident data were obtained for each site. This included any available information about incidents involving (1) near-misses, (2) medical treatment, and (3) injuries that were reportable to MSHA or the Occupational Safety and Health Administration. Incident data were summarized using spreadsheet software. The summary organized the incident data by job classification and accident type (e.g., slip/fall, caught between/struck by, sprains/strains). Incident rates by job classification were prepared for each site.

Nordic Questionnaire

Site-wide employee input was requested via response to a general discomfort questionnaire—the Nordic Questionnaire [Kuorinka et al. 1987]. These data were gathered at three of the four sites to identify musculoskeletal aches and pains existing within the working population. The goal was to have as much of the workforce as possible fill out a questionnaire. Employees were asked to complete the questionnaire after a brief overview of the instrument and reasons for its use by a safety representative from the mine or a NIOSH team member. The Nordic Questionnaire can be found in appendix B.

Supervisor Interview Guide

The NIOSH team prepared a set of interview questions to obtain relevant information from first-line supervisors. These interviews, conducted at three of the four sites, took about 30 minutes to administer. The purpose was to learn about the variety, scope, and context of work performed at the mine site and to obtain insights about the jobs and work tasks that the front-line supervisors believed to be the most physically demanding. The supervisor interview guide can be found in appendix C.

Management Concerns

As part of the field work, management[9] was asked to identify work activities that they thought warranted further study to reduce MSD injury risk.

JOB SELECTION

Baseline data were used to identify work activities considered for further study. Several factors were considered in selecting a task, but a key factor was evidence of significant MSD risk factor exposure. Tasks selected for additional study did not necessarily have the highest (within each site) risk factor exposure. The extent that each type of data (i.e., the incident data, Nordic Questionnaire, supervisor interviews, and management concerns) was used varied based on the constraints encountered at each site. However, across all four sites, the mine's incident data were thoroughly examined and management concerns were clearly identified.

A decision matrix was used at two of the sites to organize the baseline data and offer information to help select tasks for further study. A discussion of how the decision matrix was used is included in appendix E.

---

[8]Three primary criteria were chosen: (1) mine cooperation – management and workforce representatives should be interested and committed to this work; (2) mining environment – a variety of surface and underground environments should be examined; and (3) industry impact – the activities evaluated should be performed by a number of workers or have characteristics common to activities performed by a large number of workers at other mine sites or in other industries.

[9]Site employees (management and workforce representatives) collaborated in this study. Thus, discussions of management concerns were ongoing, especially during the early parts of the site-specific work where the goal was to come up with a logical means for selecting work tasks for more in-depth study.

TASK ANALYSIS

Specific tasks that exposed workers to significant MSD risk factors were identified for each candidate job. Using job descriptions acquired from supervisor interviews and discussions with mine managers, NIOSH researchers interviewed, observed, and videotaped work tasks. Team members observed and identified a set of risk factors along with relevant measurements of worksite dimensions, force exertions, and the size and weights of tools and work pieces. Finally, workers' views of their physical job stress were obtained using a Body Part Discomfort Interview Guide developed by the evaluation team.

BODY PART DISCOMFORT INTERVIEW GUIDE

Workers were interviewed to identify symptoms of discomfort and work activities that contribute to discomfort. The interview guide inquired about body part discomfort and workers' thoughts about the most physically demanding aspects of their work. The interview guide was based on concepts from the general form of the Nordic Questionnaire and a symptom survey presented as tray 4–A in NIOSH's *Elements of Ergonomics Programs: A Primer Based on Workplace Evaluations of Musculoskeletal Disorders* [Cohen et al. 1997]. The goal was to identify tasks that workers believed were most likely to cause physical discomfort and why they are hard to perform. The Body Part Discomfort Interview Guide can be found in appendix D.

Task analysis results were used by the NIOSH evaluation team to identify a list of demanding tasks. Following reviews with mine management, target tasks were chosen for final evaluation to develop ideas for improvement. A target task is a work activity with evidence of significant potential for MSD risk factor exposure. Videotapes were made for each target task to aid in the analysis and brainstorming for ideas to improve how the task is performed.

BRAINSTORMING

The NIOSH team examined all available data to identify risk factors. This information was presented at brainstorming sessions to members of the mine work groups and technical staff. The aim was to discuss targeted work activities and identify health and safety issues, with emphasis on understanding the risk factors and discussing practical ideas to reduce risk factors in order to make the activities less physically demanding.

Following the mine brainstorming sessions, a session was held with a diverse group of PRL researchers. Videotapes were reviewed by the NIOSH team to trigger a discussion of possible risk factors and to review the findings of the mine site evaluation team. Special emphasis was placed on notes taken from the brainstorming sessions conducted at the mine site. The aim was to examine the key issues identified and further develop or add to the ideas to improve how the target work tasks are performed. The risk factors identified and general ideas for job improvement were discussed at the closeout meeting held with representatives of each mine.

Ideas for improvement were presented as a listing of all ideas generated during brainstorming sessions and discussions with mine representatives. These lists were not presented in any particular order with regard to likelihood of successfully reducing risk factors or ease of implementation.

RESULTS

Appendices F through I provide detailed results for each of the four mine sites, respectively. This section provides a set of tables summarizing findings within and across each site. The tables focus on the seven common MSD risk factors discussed earlier.

SELECTING WORK ACTIVITIES AND TASKS

Table 1 describes the cooperating mine sites, baseline data used to profile MSD risk, and specific work activities/tasks identified for further study. The mines selected and studied varied with regard to commodity type, geographic location, workforce size, and mining methods.

Work tasks were selected based on the baseline data gathered at each site. Results from the analysis of the baseline data were summarized for each site. From table 1, it should be noted that the definition of an incident and the methods of tracking incidents varied for each site. Sites reporting near-misses, property damage, and minor cuts and scrapes[10] would normally yield significantly more incidents.

MSD RISK FACTORS FOR SELECTED WORK TASKS

Once work tasks were identified, the NIOSH team interviewed those performing the tasks and made observations and measurements to identify specific MSD risk factors. These risk factors are cataloged in tables 2 through 5.

---
[10]This is a good practice as it brings early attention to potential areas of concern. Thus, small problems can remain small problems.

Table 1.—Mine descriptions, tools used to select tasks, and tasks selected across the four sites

	Mines selected			
	Underground coal	Surface copper	Surface phosphate	Underground limestone
	• Eastern United States • Room-and-pillar • 44- to 56-in seam height • 126 employees: 　118 – mine 　8 – surface	• Southwestern United States • Open-pit operation with electrowinning and electrorefining • 736 employees: 　214 – mine 　522 – processing	• Southern United States • Open-pit operation that pumps slurry to a cleaning plant • 307 employees: 　146 – mine 　161 – processing and maintenance	• Eastern United States • Underground room-and-pillar quarry with a processing plant • 43 employees: 　29 – mine 　14 – processing
BASELINE DATA TO PROFILE MSD RISK				
Incident data[1]	203 records	460 records	165 records	17 records
Nordic Questionnaire	Not used[2]	507 responses (shoulders, upper back, and neck were most often reported)	50 responses (neck, knees, and shoulders were most often reported)	40 responses (low back, neck, wrists, elbows, and hands were most often reported)
Front-line supervisor interviews	Not used[2]	55 interviews – 5 tasks rated high as very physically demanding	8 interviews – 2 tasks rated high as very physically demanding	3 interviews – 3 tasks rated high as very physically demanding
TASKS SELECTED FROM THE BASELINE DATA AND CONCERNS IDENTIFIED BY MANAGEMENT				
	• Supply delivery 　Motormen • Roof bolter 　Bolt installation 　Manual materials handling • Remote-control continuous miner operator 　Cutting coal (remotely)	• Smelter operator 　Converter turnaround 　4B–5 gun rebuilds • Tank house operator 　Bar pulling 　Banding copper sheets • Refinery operator 　Cell checking	• Dragline operator 　Operating dragline • Hydraulic pit station operator 　Operating pit station • Float crew 　Wrenching	• Quality control technician 　Gathering samples 　Testing samples • Blasting crew 　Hand scaling 　Handling ANFO and dynamite • Scaling machine operator 　Operating scaling machine

[1]Information on the severity of injuries was not available; therefore, injury risk (rates) was not calculated.
[2]The NIOSH team was very familiar with this mine, tasks, and work environment. This negated the need to interview supervisors and gather information via the Nordic Questionnaire from the general workforce.

Table 2.—Risk factors by observed tasks: underground coal mine

Risk factors	Tasks					
	Supply delivery tasks			Roof bolting tasks		Continuous miner operation
	Motormen delivering and receiving railcars	Motormen collecting empty railcars and loading refuse	Motormen unloading small units of supplies	Roof bolter – bolt installation	Roof bolter – manual materials handling	Cutting coal – remote-control continuous miner
Prolonged, awkward postures		Twisting and bending of torso more than 20°	Extended forward reaches Flexing of the neck Twisting and bending of torso more than 20° Low roof conditions require kneeling or stooping during material handling	Extended time spent in kneeling posture Awkward body postures – in particular, twisting of back and neck flexion near the back Forceful hand and arm exertions when bending bolts Repetitive extended forward reaches Repetitive control use	Lifting and carrying supplies in bent or stooped postures Twisting and turning while handling supplies Working in cramped work space	Constant load on neck and shoulders from 8-lb control box Forceful exertions when lifting and moving power cable Extensive crawling Extended time spent in kneeling or stooped posture Arm or leg abduction when positioning cable
Contact stress (pressure points)			Low roof conditions require kneeling or stooping during material handling			Constant load on neck and shoulders from 8-lb control box Extensive crawling
Jolting/Jarring	Constant jarring from riding the rails					
Vibration exposure (hand/arm)						
Highly repetitive motions				Repetitive extended forward reaches Repetitive control use		Repetitive control use

Table 3.—Risk factors by observed tasks: surface copper mine

Risk factors	Tasks				
	Smelter		Tank House		Refinery
	Smelter operator – converter turnaround	Smelter operator – 4B–5 gun rebuilds	Tank house operator – bar pulling	Tank house operator – banding copper sheets	Refinery operator – cell checking
Prolonged, awkward postures	A lot of trunk twisting while handling 4B–5 guns	Reaching for, pulling, and tossing moderate to heavy items Reaching into bins, creating high forces on the low back	Asymmetric lifting task with significant trunk twisting	Handling (lifting, carrying, and positioning) of a 40-lb banding tool	Prolonged flexing of the neck due to work with low targets Arms above shoulder height when raking high nodules
Forceful exertions (heavy or frequent lifting)	Lifting and placing 45-lb guns to and from the floor Repetitive forceful pushing and pulling with a long bar to clean out tuyere holes	Reaching into bins, creating high forces on the low back Excessive pushing and pulling forces to position bin carts	Variable amounts of force are required When working alone, forceful exertions required		Significant force required to remove larger nodules Hot environment
Forceful gripping	Extended use of unsupported jackhammer		Variable amounts of force are required		
Contact stress (pressure points)					
Jolting/jarring					
Vibration exposure (hand/arm)	Extended use of unsupported jackhammer				
Highly repetitive motions	Repetitive forceful pushing and pulling with a long bar to clean out tuyere holes	Repetitive handling of 45-point guns	Highly repetitive task – hundreds of bars per day		Highly repetitive upper-extremity exertions
Other	Hot working environment, about 115 °F				

Table 4.—Risk factors by observed tasks: surface phosphate mine

Risk factors	Tasks		
	Dragline operation	Pit station operation (water cannons)	Float crew
	Operating controls	Operating controls	Wrenching pipes
Prolonged, awkward postures	Awkward foot and ankles postures Extended time in seated posture Frequent twisting and turning of head and neck	Awkward seated postures (on stools provided) Excessive neck extension when viewing pump monitors Extended time in seated posture	Awkward postures during pipe assembly task
Forceful exertions (heavy or frequent lifting)			Handling heavy and awkward materials
Forceful gripping			
Contact stress (pressure points)		Forearm contact on edge of work stations	
Jolting/jarring			Jolting/jarring while traveling over rough terrain in the maintenance truck
Vibration exposure (hand/arm)			Hand and arm vibration from air tool usage
Highly repetitive motions	Repetitive operation of foot pedals Repetitive use of joystick controls	Repetitive/continuous use of joystick controls	Repetitive hand and wrist movements
Other	Noise in operator's compartment		Pinch-point exposure

Table 5.—Risk factors by observed tasks: underground limestone mine

Risk factors	Tasks				
	Quality control technician		Blasting crew		Scaler operator
	Gathering samples	Testing samples	Hand scaling	Handling ANFO and dynamite	Operating scaler
Prolonged, awk-awkward postures	Asymmetric load handling when carrying one bucket Ascending and descending steps while carrying loads	Twisting, turning, and bending while handling sample pans Lifting loads above shoulders to dump Pulling trays and lifting loads from below knuckle height	Extended forward reaches when cleaning face Working with arms above shoulder height Working below the feet to remove material Neck flexion when working with high and low targets Wrist flexion/extension when prying material	Bending over, twisting, and reaching to lift ANFO in 50-b bags and dynamite in 55-lb boxes Loading ANFO into the hopper requires twisting and lifting with the elbow above shoulder	The neck is flexed back when scaling the back (when scaling the back from the bench, you must lean forward to look up; the neck is not well supported)
Forceful exertions (heavy or frequent lifting)	Forceful exertions with shoulders and back while shoveling Lifting and carrying heavy loads, 40- to 70-lb buckets		Forceful exertions with arms, shoulders, and back; the 8-ft-long pry bar creates a high force moment arm	Carrying awkward and heavy loads	Pushing and pulling on controls require frequent arm and shoulder exertions
Jolting/jarring					Constant bouncing and jarring while scaling
Vibration exposure (hand/arm)					
Highly repetitive motions				Repetitive control usage	
Other	Ascending and descending steps with hands occupied	Dust in the lab when analyzing a sample	Environment can be noisy, dark, smoky, foggy, and humid		

SUMMARY

This study focused on identifying MSD risk factors using qualitative and quantitative data. The tools used to assess work tasks were effective for quickly identifying risk factors. Brainstorming sessions made workers more aware of risk factor significance.

There were some obvious similarities among the four sites. Observations of work tasks and interviews with workers performing those tasks suggested that handling heavy and awkward objects, forceful arm and shoulder exertions, and working in cumbersome postures were common to all four sites. Tasks requiring a significant amount of manual work involving upper extremity and low-back activity were found at each site. Tables 2 through 5 indicate that awkward postures and forceful exertions are the most common MSD risk factors observed within and across the four sites.

Based on the task analysis, short brainstorming sessions were held with a cross-section of mine site personnel who had a stake in the task. These sessions were very useful as they generated ideas to reduce risk factors—or make the work process less physically demanding and less difficult. The sessions also offered an opportunity for the NIOSH team to meet with workers and offer comments and feedback on their work process and methods. Discussions were aided by the use of videotapes of workers performing the target tasks. In some cases, workers were surprised at the amount and extent of bending and twisting while handling heavy objects.

This project ended with a report to mine management. It was up to the mine to follow up and consider the next step—implementation or experimentation with some of the ideas generated. Consequently, it is difficult to assess the ultimate value of the ideas generated and any impact that they might have on reducing risk factors. Logical next steps include:

1. Obtaining approval from management to further explore the ideas generated during the brainstorming. If approved by management, the task is to identify resources that would be needed to implement certain ideas to reduce the exposure level or duration of identified risk factors. Examples include (a) getting feedback from the workers on the ideas generated, (b) talking with suppliers about where parts or equipment may be purchased, (c) talking with maintenance personnel to see if a solution can be fabricated in-house, and (d) finding out how other mine sites may have dealt with similar issues.

2. Identifying measures to assess the value of the proposed intervention(s) in reducing risk factor exposure levels or duration. Examples include (a) reducing the amount of weight lifted or handled, (b) reducing the number of times a task is performed, (c) reducing the amount of time needed to complete a task, (d) reducing force needed to activate a control, and (e) reducing the time spent in an awkward posture.

There are several ways of dealing with risk factors. These include combinations of engineering changes, improved work practices, and increased use of personal protective equipment. Risk factor awareness along with a proactive culture, skills, and motivation to "problem solve" the process also offer interesting solution sets, such as job rotation, a change in routine, or the need for extra help. There is often more than one way to solve a problem, and those who do the job typically have the best understanding of the problems. This was the main motive for talking with workers and others (who had a stake in the job) at each of the sites, which led to discussion of how MSD risk could be reduced for specific tasks.

The focus for job improvement should be on changes to work practices, better tool usage, or the use of new tools. Examples include (1) copper mine – methods changes for converter turnaround to make the task less physically demanding; (2) limestone mine – more frequent bucket moves and the use of a new scaling tool to limit the level of heavy work done in preparing the face for loading explosives; and (3) phosphate mine – better organization of the tools carried on the mechanic's pickup trucks to make the task of wrenching less difficult and more predictable. In each of these examples, there should be some long-term benefit in reducing the risk of sprains and strains and "caught between/struck by" injuries. Task analysis results indicated that handling heavy and awkward objects, forceful arm and shoulder exertions, and working in awkward postures were common to the four sites. This knowledge indicated a need for assist devices to help reduce manual tasks and for workers to learn more about common ergonomic risk factors and how to lessen or avoid them.

Formal training and education (e.g., identifying risk factors) is one way to begin to integrate the science of ergonomics into worklife. The objective would be to help mining personnel become better at solving problems related to MSD injury risk. The outcome is a more skilled workforce—more knowledgeable of risk factors that can impede both production and safety. This can help with longer-term organizational goals to reduce cost and improve worker safety.

REFERENCES

Cohen AL, Gjessing CC, Fine LJ, Bernard BP, McGlothlin JD [1997]. Elements of ergonomics programs: a primer based on workplace evaluations of musculoskeletal disorders. Cincinnati, OH: U.S. Department of Health and Human Services, Public Health Service, Centers for Disease Control and Prevention, National Institute for Occupational Safety and Health, DHHS (NIOSH) Publication No. 97–117.

Cornelius KM, Turin FC, Wiehagen WJ, Gallagher S [2001]. An approach to identify jobs for ergonomic analysis. In: Proceedings of the IIE Annual Research Conference. Norcross, GA: Institute of Industrial Engineers.

Fotta B, Bockosh GR [2000]. The aging workforce: an emerging issue in the mining industry. In: Bockosh GR, Karmis M, Langton J, McCarter MK, Rowe B, eds. Proceedings of the 31st Annual Institute of Mining Health, Safety and Research. Blacksburg, VA: Virginia Polytechnic Institute and State University, Department of Mining and Minerals Engineering, pp. 33–45.

Hagberg M, Silverstein B, Wells R, Smith MJ, Hendrick HW, Carayon P, et al. [1995]. Work related musculoskeletal disorders (WMSDs): a reference book for prevention. Bristol, PA: Taylor and Francis, Inc.

Hagglund G [1998]. The effect of age, occupation and experience on performance. In: Proceedings of the Global Ergonomics Conference (Cape Town, Republic of South Africa, Sept. 8–11, 1998). Amsterdam, Netherlands: Elsevier, pp. 239–249.

Kumar S [2001]. Theories of musculoskeletal causation. Ergonomics 44(1):17–47.

Kuorinka I, Jonsson B, Kilborn A, Vinterberg H, Biering-Sorensen F, Anderson G, et al. [1987]. Standardized nordic questionnaire for the analysis of musculoskeletal symptoms. Appl Ergon 18:233–237.

Scharf T, Vaught C, Kidd P, Steiner LJ, Kowalski KM, Wiehagen WJ, et al. [2001]. Toward a typology of dynamic and hazardous work environments. Hum Ecol Risk Assess 7(7):1827–1841.

Steiner LJ, Cornelius KM, Turin FC [1999]. Predicting system interactions in the design process. Am J Ind Med Sep(Suppl 1):58–60.

Stobbe TJ, Bobick TG [1986]. Musculoskeletal injuries in underground mining. Ann ACGIH 14:71–76.

U.S. Bureau of Labor Statistics [1991]. Occupational injuries and illnesses in the United States by industry, 1989. Washington, DC: U.S. Department of Labor, Bureau of Labor Statistics, Bulletin 2379.

Washington State Department of Labor [1994]. Fitting the job to the worker: an ergonomics program guideline. Olympia, WA: State of Washington Department of Labor, Division of Labor and Industries, Division of Consultation and Compliance Services, Workplace Consultation Program.

Westgaard RH, Winkel J [1997]. Ergonomic intervention research for improved musculoskeletal health: a critical review. Int J Ind Ergon 20:463–500.

Winn F, Biersner R, Morrissey S [1996]. Exposure probabilities to ergonomic hazards among miners. Int J Ind Ergon 18:417–422.

Zhuang Z, Groce D [1995]. The national occupational health survey of mining: magnitude of potential exposures to musculoskeletal overload conditions. In: Bittner AC, Champney PC, eds. Advances in Industrial Ergonomics and Safety VII. London: Taylor and Francis, pp. 273–280.

APPENDIX A.—ERGONOMIC INTERVENTIONS FOR THE MINING INDUSTRY: AN OVERVIEW OF PLANNED APPROACH

What: A NIOSH collaborative field research project to study ergonomics risk factors at a variety of mining operations.

Objectives: (1) Assess ergonomic risk factors with emphasis on musculoskeletal disorders (MSDs).

(2) Evaluate the effectiveness of interventions in terms of lowering risk factor exposures and reducing injuries.

Cooperators: The most important cooperators will be participating mine companies. MSHA has a nonregulatory, technical interest in this project. MSHA's interests include (1) learning more about MSD risk factors that mine workers are exposed to, (2) identifying the types of intervention strategies that are appropriate for mining environments, and (3) disseminating the general findings of this research throughout the mining industry. However, MSHA personnel will not be taking part in the collection and analysis of the data for this study.

Why: The dynamic nature of mining environments exposes workers to a number of injury risk factors. Commonly encountered risk factors that can lead to the development of MSDs include heavy lifting, repetitive materials handling, awkward postures, and uneven walking surfaces. There is evidence that ergonomics interventions can help to lower the occurrence of MSDs in a variety of work settings. According to a 1997 U.S. General Accounting Office study of five private sector ergonomics programs, the reduction in the number of reported injuries and illnesses ranged from 2.4 to 6.1 per 100 full-time workers.

How: A two-phase effort in which objective 1 will be addressed in phase 1 and objectives 1 and 2 will be addressed in phase 2.

Phase 1 will be concerned with gathering MSD health and safety baseline data and identifying general ideas for reducing risk factors. The goal is to visit four mine sites: underground coal (lower seam), underground nonmetal, surface metal, and surface nonmetal. At each mine, the following protocol is planned:

(1) Hold introductory meeting with management and workforce representatives.
(2) Establish safety and health baseline data using mine injury records, a musculoskeletal discomfort questionnaire, and supervisor interviews.
(3) Use baseline data to develop risk factor assessment plan.
(4) Collect MSD exposure data via worker interviews and job analysis.
(5) Analyze and interpret MSD exposure data.
(6) Prepare report of findings that identifies general ways to reduce risk factor exposures.
(7) Conduct final meeting to present findings to management and workforce representatives.

Expected phase 1 duration: four mine visits within a 6- to 12-week period.

Phase 2 will involve development, implementation, and evaluation of mine-specific ergonomics interventions. It will require long-term cooperation of one or two mine sites, which may or may not be sites that participated in phase 1. Expected duration: 3 to 4 years with many mine site visits.

Protocol:

(1) Conducted introductory meeting and site walk-through with management and workforce representatives.

(2) Established safety and health baseline data using MSHA 30 CFR 50 data, mine injury records, and a general discomfort questionnaire.

(3) Used baseline data to develop exposure assessment plans:

   (a) Prioritized jobs based on evidence of risk.
   (b) Chose jobs with significant risk for further analysis.
   (c) Selected tools for MSD exposure measurement.

(4) Collected exposure data via mine site interviews and task analysis.

(5) Analyzed and interpreted MSD exposure data.

(6) Prepared report of findings that identified risk factors and general ideas to reduce risk factor exposures at that particular mine site.

(7) Conducted final meeting to present findings to management and workforce representatives.

APPENDIX B.—NORDIC QUESTIONNAIRE

NIOSH Ergonomics Initiative
Musculoskeletal Discomfort Survey

The purpose of the survey:

You have been selected to participate in a National Institute for Occupational Safety and Health (NIOSH) study of musculoskeletal pain in the mining industry. We are hoping to determine what musculoskeletal risk factors exist at this mine and across the mining industry. We are interested in the type and location of the pain/discomfort symptoms you may be experiencing. This survey should take no more than 5 minutes to complete. Thank you for your cooperation.

How to answer the questionnaire:

Picture: In this picture you can see the approximate position of the parts of the body referred to in the questionnaire. Limits are not sharply defined, and certain parts overlap. You should decide for yourself in which part you have or have had your trouble (if any).

Table: Please answer by putting an "X" in the appropriate box - one "X" for each question. You may be in doubt as to how to answer, but please do your best anyway. Note that column 1 of the questionnaire is to be answered even if you have never had trouble in any part of your body, colullll1s 2 and 3 are to be answered if you answered yes in column 1.

Mine: _____
Initial of first name: _____ Initial of last name: _____ Last 4 digits of social security number: _____ Immediate Supervisor: _____ Date: ___/___/___
Job Title: _____ Section: _____ Gender: M F Age: _____ Height: _____ ft. _____ in. Weight: _____
How long have you been doing this job? _____ years _____ months   On average, how many hours do you work each week? _____

To be answered by everyone	To be answered by those who have had trouble	
Have you at any time during the **last 12 months** had trouble (ache, pain, discomfort, numbness) in:	Have you at any time during the **last 12 months** been prevented from doing your normal work (at home or away from home) because of the trouble?	Have you had trouble at any time during the **last 7 days**?
Neck ☐ No ☐ Yes	☐ No ☐ Yes	☐ No ☐ Yes
Shoulders ☐ No ☐ Yes, right shoulder ☐ Yes, left shoulder ☐ Yes, both shoulders	☐ No ☐ Yes	☐ No ☐ Yes
Elbows ☐ No ☐ Yes, right elbow ☐ Yes, left elbow ☐ Yes, both elbows	☐ No ☐ Yes	☐ No ☐ Yes
Wrists/Hands ☐ No ☐ Yes, right wrist/hand ☐ Yes, left wrist/hand ☐ Yes, both wrists/hands	☐ No ☐ Yes	☐ No ☐ Yes
Upper Back ☐ No ☐ Yes	☐ No ☐ Yes	☐ No ☐ Yes
Lower Back (small of back) ☐ No ☐ Yes	☐ No ☐ Yes	☐ No ☐ Yes
One or Both Hips/Thighs ☐ No ☐ Yes	☐ No ☐ Yes	☐ No ☐ Yes
One or Both Knees ☐ No ☐ Yes	☐ No ☐ Yes	☐ No ☐ Yes
One or Both Ankles/Feet ☐ No ☐ Yes	☐ No ☐ Yes	☐ No ☐ Yes

*Based on the Nordic Questionnaire

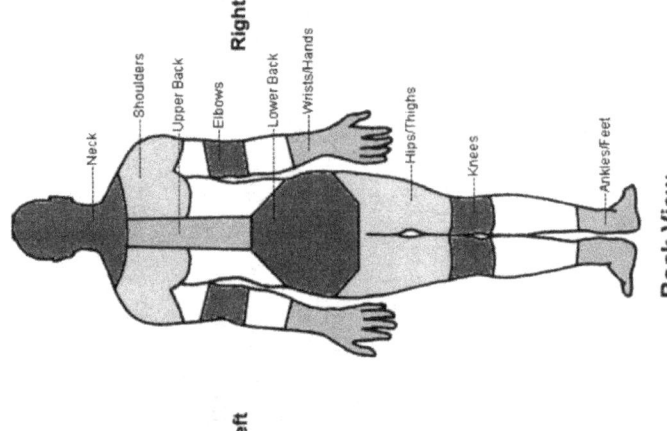

APPENDIX C.—FRONT-LINE SUPERVISOR INTERVIEW GUIDE

I am part of a NIOSH research team that is working on a small study to evaluating physical risk factors at this mine. In particular, we want to obtain information about the causes of musculoskeletal pain. What we mean by "musculoskeletal pain" is pain that may result in injuries that are commonly referred to as "sprains" or "strains."

I would like to spend a few minutes to interview you about work activities in the area or group that you supervise. Our team plans to select three or four work activities for a final analysis, and we need your help. This interview should take about 20 minutes.

Mine: ___   Interviewer:_____   Date ___/___/___

Initial of first name: __   **Initial of last name**: __
Last 4 digits of social security number: __ __ __ __

Title:_____   **Work location**:_____

How long have you been doing this job? _____years _____months

Shift information:

Shifts worked in this area: _____   **Do you rotate shifts?** Y N

If yes, describe the shift rotation sequence:

Supervisory responsibility: **Number of employees supervised:** _____

Are these employees part of a work crew (like a production crew)? Y   N

Do you supervise the same people day in and day out? Y   N

Do you supervise the same job classifications day in and day out? Y   N

Information about the work crew/group:
Please describe the jobs, title, and brief description performed by your work crew/group:
1.

For these jobs, which ones seem to be the most physically demanding? Why?

Are there any tasks or activities within these jobs that are especially difficult or physically demanding?
(Probe for specific factors – environmental factors like heat, humidity, dust, noise ... Or awkward postures ... Or handling heavy loads... Or types of tools or equipment used ... Or repetitious work.... Or pressures to get the job done... etc.)

Is there anything else about the work done in your area that you think is relevant?

APPENDIX D.—BODY PART DISCOMFORT INTERVIEW GUIDE

Mine: _____ Interviewer: _____ Date: __/__/__ Target Task: _____ Unique Identifier: _____

Job Title: _____ Section: _____ Shift: _____ Average # of hours worked per week: _____

Age: _____ Gender: M F Height: _____ Weight: _____ Env Conditions (task): _____

Immed. Supervisor: _____ How long have you worked at this facility? _____ yrs _____ months …at this particular job? _____ yrs _____ months

Body Part Discomfort Interview

As a result of doing this job, have you experienced discomfort or pain **within the past year** in your:

Body Part	Freq.	Sev.	Related Work Activities	Comments (Describe Pain/Treatments)
Neck	1 2 3 4	1 2 3 4		
Shoulders	1 2 3 4	1 2 3 4		
Elbows	1 2 3 4	1 2 3 4		
Wrists	1 2 3 4	1 2 3 4		
Hands	1 2 3 4	1 2 3 4		
Upper Back	1 2 3 4	1 2 3 4		
Mid Back	1 2 3 4	1 2 3 4		
Lower Back	1 2 3 4	1 2 3 4		
Upper Legs	1 2 3 4	1 2 3 4		
Knees	1 2 3 4	1 2 3 4		
Lower Legs	1 2 3 4	1 2 3 4		
	1 2 3 4	1 2 3 4		

Frequency: (1) 1-2 Times/Year (2) 1-2 Times/Month (3) 1-2 Times/Week (4) Every Day

Severity: (1) Mild pain or discomfort (2) Moderate pain with no reduction in activity (3) Severe pain with reduction in activity (4) Unbearable pain requiring time off work

Please shade in area(s) of discomfort
(Indicate **F**ront or **B**ack when appropriate)

Left — Shoulders, Upper Arm, Elbows, Lower Arm, Hand

Right — Neck, Upper Back, Mid Back, Lower Back, Buttocks, Upper Leg, Knees, Lower Leg

Back View

What are the most physically demanding aspects of this job?

What improvements would you like to see for this job?

Interview Notes:

APPENDIX E.—SAMPLE DECISION MATRIX

Mine incident data, Nordic Questionnaire responses, and supervisor interviews were used to construct a decision matrix for identifying work groups appropriate for further evaluation. For each source of data, a subjective ranking of "low," "medium," or "high" was given by consensus of the NIOSH evaluation team. Each ranking for a work group was relative to other groups at the site. A ranking of "high" would not indicate a high risk level for injury; it would indicate that the comparison measure(s) for the given data type was high when compared to other work groups at that site.

For each work group, an incident ratio was calculated to provide a relative estimate of incident risk. A work group incident ratio significantly above the site-wide incident ratio was given a "high" rating. A work group incident ratio significantly below the site-wide incident ratio was given a "low" rating.

Work group discomfort was evaluated by comparing the number of body parts reported and the number of reports of discomfort for similar body parts. Work groups reporting higher than the site-wide average numbers of body parts and high percentages of workers with discomfort for a body part(s) were given a "high" rating. Work groups reporting lower than the site-wide average numbers of body parts and low percentages of workers with discomfort for a body part(s) were given a "low" rating.

If a supervisor identified a work group as having physically demanding work, it was considered for a "high" or "medium" rating. Supervisor comments were evaluated for key characteristics of physical stress. The degree of physical stress was based on common physical risk factors, which include forceful exertions, heavy lifting, awkward postures, repetitive motions, and jolting or jarring. If any single or combination of risk factors was described to exist at a significant level, a rating of "high" was given. Work groups not identified by a supervisor were given a "low" rating.

For scoring purposes, a rating of "high" was given 3 points, "medium" was given 2 points, and "low" was given 1 point. One additional point was awarded if management expressed concern regarding a particular work group. Work groups having the highest scores were deemed suitable for further evaluation. Cornelius et al. [2001] provide a more detailed discussion of how the decision matrix can be used to identify jobs that could be suitable for ergonomic evaluation. Table E–1 shows a sample decision matrix.

Table E–1.—Sample decision matrix

Work group	No. of employees	Incident data	Nordic Questionnaire	Supervisor interviews	Management concern	Final score
Supervisors	4	Low	High	Low		5
Haul truck operators	15	Medium	Medium	Low	Yes	6
Driller operators	5	Medium	Low	Low		4
Blasting crew	4	High	High	High	Yes	10
High-lift/loader operators	4	Medium	Medium	Medium		6
Mechanics	4	High	High	High		9
Water truck operator	1	Low	High	Low		5
Laborers	2	Low	High	High	Yes	8
Welders	3	Low	Medium	Medium		4

APPENDIX F.—UNDERGROUND COAL MINE

This mine is an underground bituminous coal operation in the Eastern United States with 126 employees. Work activities include advance-and-retreat continuous mining using mainly shuttle cars and remote-control continuous mining machines. Seam height ranged from 44 to 56 in.

The evaluation plan consisted of four steps. The first step was to use mine injury records and a list of management concerns to select work groups that had significant evidence of MSD risk factors. The NIOSH evaluation team members were very familiar with underground coal mine work activities, and several members were familiar with this particular mine site. For this reason, supervisor interviews and Nordic Questionnaires were not used. The five work groups chosen for evaluation were supply/delivery personnel, roof bolting machine operators, continuous mining machine (CMM) operators, shuttle car operators, and foremen/mine examiners.

The second step was to use work task observation data, comments from the mine management team, and responses to the Body Part Discomfort Interview Guide to select target tasks for final evaluation. Supply motormen had three target tasks: delivering railcars to section side tracks, unloading supplies from railcars, and loading refuse onto motor cars. The target task for the supply yard was the retrieval of garbage and recyclable items from railcars by high-lift operators. Target tasks for roof bolter operators were installing roof bolts and retrieving supplies. The target task for CMM operators was using the machine to cut coal. The target task for shuttle car operators was transporting coal. Target tasks for foreman/mine examiners were walking throughout the mine and materials handling.

The third step was to use tailored task analysis to characterize the ergonomic risk factors and musculoskeletal symptoms that exist for target tasks. The final step was to discuss the ergonomic risk factors for each target task with mine management and the workers who perform the tasks to identify ways to reduce risk factor exposures.

INCIDENT DATA

The mine provided 3 years of incident data from January 1, 1996, to December 31, 1998. These included (1) lost-time injuries, (2) medical treatment, and (3) workmen's compensation claim data. Incident data were compiled and summarized by the NIOSH team. A total of 203 records were obtained. Data were categorized by occupation. Table F-1 summarizes these incidents by occupational class. CMM operators and scoop operators had the highest reported incident rate.

Table F–2 summarizes the 203 incidents by occupational class and incident type. Each incident record was reviewed and classified as one of six types: (1) slip/fall, (2) caught between/struck by, (3) strain/sprain, (4) burn, (5) other, or (6) unknown. Strain/sprain was split into subtypes of overexertion, awkward posture, and jarred/bounced/vibrated. The two most common incident types were "caught between/struck by" and strain/sprain. They were common to each of the occupational classes. Of the 203 reported incidents, 92 (45%) were "caught between/struck by" incidents and 77 (38%) were sprains/strains.

TASK-SPECIFIC RESULTS

Tables F–3 to F–6 summarize information gathered for the specific tasks studied at the underground coal mine. Ideas to reduce risk factors were developed based on brainstorming sessions with those who had a stake in the task, i.e., those who performed the task, supervisors, safety representatives, and engineering and maintenance staff. NIOSH personnel facilitated the brainstorming sessions. It should be noted that all ideas generated during brainstorming are listed. The ideas for improvement are not listed in any particular order with regard to likelihood of successfully reducing risk factor exposure or ease of implementation. These ideas are unique to this mine site and the tasks observed and should not be generalized to all mine sites.

Table F–1.—Underground coal: number of incidents and employees by job title and incident rate, 1996–1998

Occupation	Incidents	Employees	Average annual incident rate[1]
Continuous miner operators	18	6	1.0
Scoop operators	18	6	1.0
Supply clerks	7	4	0.58
Foremen	52	30	0.58
Mechanics	23	16	0.48
Roof bolter operators	34	24	0.48
Shuttle car operators	22	18	0.47
General laborers	9	22	0.14
Unknown occupations	20	NA	NA
Total	203	126	NA

NA Not available.
[1]Average annual incident rate = reported incidents divided by estimated number of employees divided by 3 years.

Table F-2.—Underground coal: incidents by occupational class and incident type, 1996–1998

Incident type	Continuous mining machine operators	Foremen	Supply and laborers	Roof bolter operators	Scoop and shuttle car operators	Mechanics	Unknown	Total
Slip/fall		8	1	2	2	1	3	17
Caught between and struck by	10	23	9	12	13	11	14	92
Strain/sprain:								
Overexertion	3	8	4	6	14	8	2	45
Awkward posture	2	8	1	11	4			26
Jarred/bounced/vibrated		3			3			6
Burn			1	1		1		3
Other	3	1		2	4	2	1	13
Unknown		1						1
Total	18	52	16	34	40	23	20	203

Table F-3.—Underground coal: roof bolter operator – installing bolts and manual materials handling

Target tasks	Primary risk factors
Installing roof bolts (N=24) Operators drill holes in the mine roof and install bolts.	• Awkward body postures, in particular: twisting of back, flexing the neck, and extended time spent in kneeling posture • Forceful work – hand and arm exertions when bending bolts • Repetitive extended forward reaches • Repetitive control use
Manual materials handling (N=24) Operators must load and handle cable, resin, bolts, wrenches, and plates.	• Lifting and carrying supplies in bent or stooped postures • Twisting and turning while handling supplies • Forceful exertions when handling power cable • Extended time spent in awkward and cramped work space

Part of the body affected:

Discomfort surveys not done. From NIOSH observations of risk factors: back, shoulders, neck, knees, elbows, and wrists.

Ideas for improvement

- Use effective knee padding. Knee pads should allow for mobility; protect the knees from hard, uneven surfaces; and provide cushioning (such as a gel lining) where it contacts the knee cap.
- Consider using folding seats attached to the roof bolting machine to get workers off of their knees and give them the opportunity to vary their posture.
- Use job rotation, which would prevent these workers from constantly being on their knees.
- Better organize tool and supply trays on the bolting machines. It is important to ensure that there are no barriers in the path of handling materials.
- Redesign control placement to improve operator positioning.
- To aid in bolt bending, use a pipe mounted on the machine, which can be used to get leverage by placing one end of the bolt in the pipe.
- Ensure that supplies are stored close to the machine and in manageable weights.
- Supplies should be stored on the machine in an orderly fashion. This decreases the force required to pull out supplies from the bottom of a pile or to lift heavier materials to get to something else.

Table F–4.—Underground coal: continuous mine operator – cutting coal

Target tasks	Primary risk factors
Cutting coal (N=6) Miner operators were responsible for remotely operating the machine that cuts coal and loads it into shuttle cars.	• Contact stress – constant load on neck and shoulders from 8- b control box • Forceful exertions when lifting and moving power cable • Awkward postures – extensive crawling, extended time spent in kneeling or stooped posture, and arm or leg abduction when positioning cable • Repetitive control use

Part of the body affected:
Discomfort surveys not done. From NIOSH observations of risk factors: back, neck, and knees.

Ideas for improvement
• Use a support device to stabilize control box and remove the weight from the neck. Operators can use several devices to help support the control box. An example is short chains (3–4 in) that can be hooked onto the miners' belt and onto the remote-control box. These chains allow the operator to "rest" the remote box by hanging the weight from the belt. Another support device is a "pogo stick." It is a single-legged pole (pole height can be changed for different seam heights) that screws onto the bottom of the remote box. The stick is balanced by the operator. • Use rugged knee pads with soft gel inner linings to help alleviate knee pain from kneeling. • Provide operators with a seat or stool to operate from to reduce knee and low back stress. Several continuous mining machine operators have used buckets with cushions to sit on. • Consider a reel device on the continuous miner to help position cable. • Get help when pulling on long lengths of cable.

Table F–5.—Underground coal: shuttle car operator – transporting coal

Target task	Primary risk factors
Transporting coal (N=18) Shuttle car operators are respons ble for hauling coal from the face area to the belt for transfer outside.	• Considerable vibration and jarring • Awkward postures and cramped spaces • Forceful exertions when handling power cable

Part of the body affected:
Discomfort surveys not done. From NIOSH observations of risk factors: back and neck.

Ideas for improvement
• Replace the current foam-padded seating on shuttle cars with seats that use lab- and field-tested viscoelastic foams.

Table F–6.—Underground coal: motormen – unloading supplies

Target tasks	Primary risk factors
Motormen unloading small units of supplies (N=6) Most items are unloaded by scoop operators, but items such as cap wedge bundles and partial rock dust pallets are manually unloaded.	• Awkward postures: flexing of the neck, extended forward reaches, twisting and bending of torso more than 20°, and low roof conditions require kneeling or stooping during material handling

Part of the body affected:
Discomfort surveys not done. From NIOSH observations of risk factors: neck, shoulders, and back.

Ideas for improvement
• The manual handling of supplies should be reduced as much as poss ble. Aim to send all items underground on pallets. If there are materials currently delivered by the supplier in bundles that are too large to go underground, then work with suppliers so that acceptable bundle sizes can be acquired. Try to use frequent and open communication and more underground storing to avoid sending partial pallets underground. • Because communication is a problem, try to develop a better supply-tracking system. This would involve clearly defining the information needs of everyone involved in the supply delivery and receipt process. Then clearly establish a responsibility chain for keeping the tracking system up to date so that everyone in the process has current and relevant information.

APPENDIX G.—SURFACE COPPER MINE

This mine is a surface copper operation in the southwestern United States with more than 700 employees. The extensive operations include an open pit for copper ore extraction, leaching pads and ponds for collecting copper in solution, an electrowinning plant for plating nearly pure copper cathodes from solution, a smelting plant for producing copper anodes, an electrorefinery for breaking down copper anodes from the smelter into solution and plating nearly pure copper cathodes, and a rod plant to produce coils of copper rod.

The evaluation plan consisted of four steps. The first step was to use mine incident records, the Nordic Questionnaire, supervisor interviews, and a list of management concerns to select work areas that have significant evidence of MSD risk factors. A decision matrix was developed using the results of an evaluation of the data. The three work areas selected for further evaluation were smelter operations, refinery operations, and tank house operations. The second step was to use work task observation data and comments from the mine management team to select target tasks for final evaluation. The four target tasks were converter turnaround and 4B-5 gun rebuilds at the smelter, bar pulling and banding at the tank house, and cell checking at the refinery. The third step was to use body part discomfort interviews and tailored task analysis to characterize the ergonomic risk factors and musculoskeletal symptoms that exist for target tasks. The final step was to conduct brainstorming sessions with workers who perform target tasks to discuss the ergonomic risk factors and to identify general ideas to reduce risk factor exposures.

A brief overview of the baseline data, a copy of the decision matrix, risk factors identified, and general ideas for task improvement are presented below.

INCIDENT DATA

The mine provided 3 years of incident data from January 1, 1996, to December 31, 1998. These incidents involved (1) near-misses, (2) medical treatment, and (3) lost work time. Incident data were summarized by the evaluation team. A total of 475 records were obtained.

Data were categorized by work area. To determine which work areas had higher incident rates, researchers obtained an estimated number of employees for each work area. Table G-1 summarizes these incident rates. Incident rates for areas such as administration, land, and the warehouse were not calculated due to the small number of employees for each, but the injury types for these areas are included in table G-2. The highest reported incident rates occurred at smelter operations, smelter maintenance, and rod mill operations. Table G-2 summarizes the 475 incidents by work area and incident type. Each incident record was reviewed and classified as one of six types: (1) burn, (2) slip/fall, (3) caught between/struck by, (4) strain/sprain, (5) other, or (6) unknown. The two most common incident types were "caught between/struck by" and strain/sprain. Of the

Table G-1.—Surface copper: number of incidents and employees by work area and incident rate, 1996-1998

Work area	Incidents	Employees	Average annual incident rate[1]
Mine operations	70	144	0.16
Mine maintenance	30	70	0.14
Refinery operations	32	73	0.15
Refinery maintenance	7	22	0.11
Smelter operations	144	153	0.31
Smelter maintenance	87	96	0.30
Rod mill operations	32	40	0.27
Rod mill maintenance	3	12	0.08
Solution extraction (SX) operations	22	43	0.17
Solution extraction (SX) maintenance	9	21	0.14
Tank house operations	13	39	0.11
Tank house maintenance	11	23	0.16
Total	460	736	0.21

[1]Average annual incident rate = reported incidents divided by estimated number of employees divided by 3 years.

475 incidents, 184 (39%) were "caught between/struck by" incidents. Nearly half of the "caught between/struck by" incidents occurred at the smelter. Workers in the smelter and the mine areas were found to have the most sprains and strains. Most strain/sprain injuries and many "caught between/struck by" injuries are indicators of MSD risk factors.

NORDIC QUESTIONNAIRE

Table G-3 shows body part discomfort data rated as "above average" or "below average" by work area. To be considered "above average," ratings had to be in the upper third of all ratings. Similarly, "below average" ratings were in the lower third of all body part discomfort ratings for a specific body part. If a body part does not appear in the table, it can be considered to have an "average" rating of discomfort.

SUPERVISOR INTERVIEWS

Fifty-one jobs were identified by supervisors as physically demanding—17 maintenance jobs and 34 operations jobs. Of the 34 operations jobs, 5 were given a "low" exposure rating, 5 were given a "high" exposure rating, and 24 were given a "medium" rating. The five "high"-rated jobs were cell checking in the refinery, ISA tapping, converter turnarounds, smelter technicians, and bar pullers in the tank house.

Of particular note are the exposure ratings for maintenance jobs. These jobs were the most difficult to rate. It was decided to rate them all as "medium." The rationale was that maintenance tasks may vary significantly each day. For each maintenance job identified, there was at least one associated task that could have been rated "high." Because this evaluation was to be for jobs as a whole, it was believed to be best to rate them all as "medium."

G-2.—Surface copper: incidents by work group and incident type, 1996-1998

Work group	Incident type						Total
	Burn	Slip/fall	Caught between/ struck by	Strain/ sprain	Other	Unknown	
Administration			4	1			5
Land		2					2
Mine maintenance		7	14	8	1		30
Mine operators		9	19	21	9	13	70
Project services		1					1
Quality control lab	1	1					2
Refinery maintenance	1	2	3			1	7
Refinery operators	3	3	15	10	1		32
Rod plant maintenance	1		1	1			3
Rod plant operators	5	1	18	5	2	1	32
Sample plant operators			1				1
Smelter maintenance	8	10	37	21	3	8	87
Smelter operators	28	23	51	24	5	13	144
Solution extraction (SX) maintenance	2	2	3	2			9
Solution extraction (SX) operators	2	6	10	4			22
Tank house maintenance	1	4	2	3	1		11
Tank house operators	1	2	5	3		2	13
Warehouse			1				1
Other		3					3
Total	53	76	184	103	21	38	475

Table G-3.—Surface copper: ratings of body part discomfort by work area

Work area	Above average ratings	Below average ratings
Smelter maintenance	Knees, ankles	Neck, shoulder, low back.
Smelter operators	None	None.
Mine maintenance	Hips	Neck, wrist, low back, knee, ankle.
Mine operators	Upper back	None.
Rod plant maintenance	None	All.
Rod plant operators	Shoulders, e bows	All.
Tank house maintenance	Shoulders, low back	E bows, hips, knees, ankles.
Tank house operators	Shoulders, e bows, upper back, knees, ankles	None.
Solution extraction (SX) maintenance	Hips	Low back, neck.
Solution extraction (SX) operators	Neck, shoulders, upper back, hips, ankles	None.
Refinery maintenance	Neck, shoulders, hips, knees	None.
Refinery operators	Neck, upper back	None.

DECISION MATRIX

Results of the decision matrix are shown in table G-4. The ranking procedure identified three areas with a score of 8 or higher: refinery operators (9), smelter operators (8), and tank house operators (8).

BODY PART DISCOMFORT INTERVIEWS

Body part discomfort interviews were conducted with 19 workers who perform target tasks. The parts of the body most cited by converter turnaround workers were shoulder, elbow, wrist, midback, and low back. The average severity of reported pain was mild to moderate, but the frequency of pain was weekly or daily. This indicates that the pain is generally not too severe, but is experienced frequently in many parts of the body. It should be noted that no more than 5 out of the 11 converter turnaround workers reported pain in any one area of the body. Similarly, only one of three bar pullers reported mild, but frequent pain in the neck and low back. All four cell checkers reported pain in the neck, elbow, and low back. The most severe pain was reported for the elbow.

TASK-SPECIFIC RESULTS

Tables G-5 to G-8 summarize information gathered for the specific tasks studied at the surface copper mine. Ideas to reduce risk factors were developed based on brainstorming sessions with those who had a stake in the task, i.e., those who performed the task, supervisors, safety representatives, and engineering and maintenance staff. NIOSH personnel facilitated the brainstorming sessions. It should be noted that all ideas generated during brainstorming are listed. The ideas for improvement are not listed in any particular order with regard to likelihood of successfully reducing risk factor exposure or ease of implementation. These ideas are unique to this mine site and the tasks observed and should not be generalized to all mine sites.

Table G–4.—Surface copper: decision matrix for selecting tasks

Work area	No. of employees	Incident data	Nordic Questionnaire	Supervisor interviews	Management concern	Final score
Mine:						
Maintenance	70 (10%)	Medium	Low	Medium		5
Operators	144 (20%)	Medium	Medium	Medium		6
Refinery:						
Maintenance	22 (03%)	Low	High	Medium		6
Operators	73 (10%)	Medium	High	High	Yes	9
Rod plant:						
Maintenance	12 (02%)	Low	Low	Medium		4
Operators	40 (05%)	High	Low	Medium		6
Smelter:						
Maintenance	96 (13%)	High	Medium	Medium		7
Operators	153 (21%)	High	Medium	High	Yes	9
Solution extraction (SX):						
Maintenance	21 (03%)	Medium	Medium	Medium		6
Operators	43 (06%)	Medium	High	Medium		7
Tank house:						
Maintenance	23 (03%)	Medium	Medium	Medium		6
Operators	39 (05%)	Low	High	High	Yes	8

Table G–5.—Surface copper: smelter operator – converter turnaround

Target task	Risk factors
Converter turnaround (N=11) This task consisted of removing 4B–5 guns from the converter, identifying guns requiring repair, cleaning out tuyere holes, and reinstalling 4B–5 guns.	Primary risk factors: • Hot working environment, about 115 °F • Lifting and placing 45-lb guns to and from the floor • A lot of trunk twisting while carrying and placing guns • Repetitive, forceful pushing and pulling to clean out holes • Extended use of an unsupported (29-lb) jackhammer
Part of the body affected:	
Discomfort surveys were used. All 11 workers reported pain, with no more than 5 out of 11 respondents reporting pain in any particular area of the body. Average to moderate pain: shoulder, e bows, wrists, midback, and low back. Frequency of pain: weekly or daily.	
Ideas for improvement	
• Use a rack or cart to hold guns to reduce lifting, especially from the floor. The rack or cart could be positioned to reduce the distance that guns are carried. • Use a wheeled rack that can be maneuvered into place with guns, then wheeled out to rebuild. • Either suspend or support the jackhammer to reduce the amount of force required to position and hold the pneumatic hammer while cleaning the holes. • Explore replacing the jackhammer and bar with a mobile drill to clean holes. • Ensure that there is sufficient air movement to reduce nonradiant heat effects. If the source is radiant heat, control by blocking the heat source.	

Table G–6.—Surface copper: smelter operator – 4B–5 gun rebuild

Target task	Risk factors
4B–5 gun rebuilds (N=1) This task consisted of moving, checking, lubricating, testing, and sometimes rebuilding 4B–5 guns.	Primary risk factors: • Repetitive handling of 45-lb guns • Reaching into bins, creating high forces on the low back • Excessive pushing and pulling forces to position bin carts
Part of the body affected:	
Discomfort survey was used. Worker reported pain in the elbows, hands, and knees.	
Ideas for improvement	
• Use a rack/cart to hold guns so that worker can test more than one gun at a time (set up in sequence). • Put the rebuild shop in the converter aisle and deliver guns in a rack/cart ready to be tested. • If using a rack to store and work on guns, the guns should be clamped in and racks blocked when testing. • Use a ratcheting tool with a better grip to remove O-rings.	

Table G-7.—Surface copper: refinery operator – cell checking

Target task	Risk factors
Cell checking (N=4) This task required workers to identify and correct shorts in plating cells. Cell checking requires workers to locate and knock off nodules on copper cathodes using a hand rake.	Primary risk factors: • Highly repetitive upper-extremity exertions • Significant force required to remove nodules • Hot environment • Prolonged flexing of the neck due to work with low targets • Awkward shoulder postures while raking
\multicolumn{2}{Part of the body affected:}

Part of the body affected:

Discomfort survey was used. All four workers reporting pain in neck, elbow, and low back. The most severe pain was reported for the elbow.

Ideas for improvement

- Consider a new power-assisted, spring-loaded, or laser-cutting tool.
- Put a second handle on the bar to help with posture when removing higher nodules.
- Use cushioned or dampening material on handles of rakes.
- Develop a way to allow cell checkers to look forward instead of down while performing the task, e.g., use a mirror to display low targets.
- When checking for shorts, mount the display meter higher or use an audible cue and an earplug.
- Use footwear that will provide protection from the heat and offer antifatigue properties.
- Use job rotation in the refinery.

Table G-8.—Surface copper: tank house operator – bar pulling

Target task	Risk factors
Bar pulling (N=3) This task required the worker to pull hanger bars out of the loops attached to a copper cathode.	Primary risk factors: • Highly repetitive task – hundreds of bars per day • Variable amounts of force are required • When working alone, forceful exertions required • Asymmetric lifting task • Significant trunk twisting

Part of the body affected:

Discomfort surveys were used. One worker reported mild, but frequent pain in the neck and low back

Ideas for improvement

- The current bar-pulling task should be a two-person task.
- Consider using an automated bar-pulling process like that observed at another mine.
- Adjust the cathode tilt table and bar-receiving table to ensure good posture and minimal transfer distance.
- Spring load the bar-receiving table. The table cannot be bolted down. It needs to be moved by the forklift when full.
- Spray the bars with a lubricant. The bars used to be sprayed with oil, but this caused problems with the quality of the copper.
- The banding tool could be suspended and counterweighted to allow for easier handling.
- The banding tool could be mounted on a platform. The platform should be roughly the same height as the cathodes.

APPENDIX H.—SURFACE PHOSPHATE MINE

This mine is a surface phosphate operation in the Southern United States with 307 employees. The operations include a cleaning plant and five phosphate pits, each with a dragline, pit car, and dozer. Mining operations involve overburden removal, phosphate extraction, slurry pits, and pumping and piping slurry (referred to as "matrix") back to the plant. The cleaning plant removes impurities from the matrix and loads the phosphate into railcars for shipment to concentration plants.

The evaluation plan consisted of three steps. The first step was to use mine incident records, the Nordic Questionnaire, supervisor interviews, and a list of management concerns to select work groups that have significant evidence of MSD risk factors. The evaluation team only had time to gather data for the following work groups: float crews, pit operators, dragline operators, and reagent operators. The second step was to use work task observation data and comments from the mine management team to select target tasks for more in-depth evaluation. The target tasks selected were wrenching and truck driving by float crews, operation of the hydraulic pit station, operation of the dragline, emptying reagent cars, and general dozer operation. The final step was to use body part discomfort interviews, video task analysis, and brainstorming sessions with mine workers to characterize the ergonomic risk factors and develop general risk reduction ideas.

A brief overview of the baseline data, risk factors identified, and general ideas for task improvement are presented below.

INCIDENT DATA

The mine provided 3 years of incident data from January 1, 1996, to December 31, 1998. A total of 165 records were obtained. Because of numerous reorganizations that occurred, incident rates were calculated based on the total number of employees as of January 1, 2000, rather than average numbers over the 3-year period. Table H–1 lists job titles, number of incidents, and incident rates. Table H–2 summarizes the 165 incidents by occupational class and incident type. Each incident record was reviewed and classified as one of six types: (1) burn, (2) slip/fall, (3) caught between/struck by, (4) strain/sprain, (5) other, or (6) unknown. "Caught between/struck by" and "strain/sprain" were the incidents that occurred most frequently and accounted for over 50% of the total.

NORDIC QUESTIONNAIRE

Table H–3 shows Nordic Questionnaire responses rated as "above average" or "below average" by work group. To be considered "above average," response percentages had to be at least 50% higher than the mine-wide percentage for reported discomfort for either the past year or the past week. Similarly, "below average" response percentages had to be at least 50% lower than the mine-wide percentage for reported discomfort for

Table H–1.—Surface phosphate: number of incidents and employees by job title and incident rate, 1996–1998

Job title	Incidents	Employees	Average annual incident rate[1]
Cleaning plant operators	22	58	0.13
Crane operators	4	4	0.33
Dozer operators	2	13	0.05
Dragline operators	4	20	0.07
Electricians	5	18	0.09
Environmental technicians	1	8	0.04
Laborers	7	43	0.05
Mechanics/repairmen, mechanic helpers, welders	54	45	0.40
Pit operators	1	16	0.02
Wrenchers, truck drivers	28	25	0.37
All other occupations	37	57	0.22
Total	165	307	0.18

[1] Average annual incident rate = reported incidents divided by estimated number of employees divided by 3 years.

either the past year or the past week. If a body part does not appear in the table, it was considered to have an "average" rating of reported discomfort. The low number of respondents (50) weakens inferences about reported body part discomfort. The results were most useful for those work groups with a larger number of respondents, i.e., the preparation plant and the pit crew.

SUPERVISOR INTERVIEWS

This mine utilizes supervisors for the float crews (float supervisor), dragline pits (area supervisor), and flotation plant (shift leader). All of these positions can be held by a regular supervisor or a "step-up" supervisor, which is a laborer who has been temporarily "stepped up" into the supervisory position. These employees have a dual role. They perform work group coordination and regular job duties. About 35% of the supervisors interviewed during all field visits were "step-up" supervisors.

Twelve jobs were identified as physically demanding—seven in the mine and five in the plant. Of the seven mine jobs, three were given a "low" exposure rating, two were given a "high" exposure rating, and two were given a "medium" exposure rating. Of the five plant jobs, three were given a "low" rating and two were given a "medium" rating. Determination of degree of physical stress was predicated on description of common physical risk factors, which include forceful exertions, awkward postures, repetitive motions, duration of exposure, contact stresses, vibration, or combinations of these factors. Table H–4 lists key responses of supervisors for jobs given a "high" or "medium" rating by the NIOSH evaluation team.

Table H–2.—Surface phosphate: incidents by job title and incident type, 1996–1998

Job title	Incident type						Total
	Burn	Slip/fall	Caught between/ struck by	Strain/ sprain	Other	Unknown	
Cleaning plant operators		8	5	2	7		22
Crane operators		1	1		2		4
Dozer operators				1	1		2
Dragline operators		3				1	4
Electricians	2		1	2			5
Environmental technicians					1		1
Laborers		2	2	2	1		7
Mechanics/repairmen, mechanic helpers, welders	3	2	24	12	13		54
Pit operators		1					1
Wrenchers, truck drivers		2	9	7	7	3	28
Other/unknown	1	5	9	10	7	5	37
Total	6	24	51	36	39	9	165

Table H–3.—Surface phosphate: ratings of body part discomfort by work area

Work group	Employees	Above average ratings	Below average ratings
Preparation plant	23	Ankles/feet, knees	E bows.
Pit crew	13	Neck, shoulders	None.
Administrative	7	Neck, elbows, lower back	Knees and upper back.
Maintenance	3	Shoulders, e bows, upper back, hips/thighs, knees.	Ankles/feet.
Other	3	None	Shoulders, elbows, wrists/hands, upper back, hips/thighs, knees, ankles/feet.
Float crew	1	None	None.

Table H–4.—Surface phosphate: physically demanding jobs identified by supervisors

Work area	Physically demanding job	Key comments	Rating
Mine	Wrencher – moves, rolls, connects, and maintains pipe, pit moves, move power cables.	Wrenching bolts is hard on back because of awkward positions and heavy tools; poor footing in muddy, uneven terrain. Exposed to pinch points.	High.
Mine	Laborer – probationary period, same tasks as wrencher.	Same as wrencher	High.
Mine	Truck driver – drives float crew truck, handles bolts and supplies.	Mentally more demanding, does some wrenching; pulling air hose is sometimes difficult.	Medium.
Mine	Pit station operator – operates hydraulic cannons.	Sits a lot, requires constant hand movements to manipulate cannons.	Medium.
Plant	Spiral operator (and assistant) – takes care of spirals.	Physical work cleaning spirals with steel wool, volume of work and number of spirals, stand on feet, constant motion.	Medium.
Plant	Waste system operator – maintains spillways	Lifting of heavy boards in and out of truck and at spillway.	Medium.

BODY PART DISCOMFORT INTERVIEWS

NIOSH researchers interviewed 26 wrenchers/truck drivers, 7 dragline operators, 3 pit operators, 4 dozer operators, and 2 reagent operators. The parts of the body with discomfort most cited by workers performing wrenching tasks were low back, wrists, and shoulders. The average pain severity was mild to moderate, and the average frequency of pain was monthly. Results for the wrenching task indicate that the pain is generally not too severe, but is experienced frequently in many parts of the body. Results for dragline operators are more dispersed, with no one body part being cited by the majority of operators. Average discomfort was mild to moderate. However, the average frequency was cited as weekly to daily. Of the four dozer operators, all reported neck and low back pain; the pain severity was on average mild, but occurred weekly. Among the three workers operating the pit station, all indicated that they had shoulder pain. The pain severity was mild to moderate and occurred almost weekly. Finally, the two reagent operators reported neck, shoulder, wrist, hands, back, and knee discomfort. This discomfort was mild to moderate, but was experienced daily.

TASK-SPECIFIC RESULTS

Tables H–5 to H–8 summarize information gathered for the specific tasks studied at the surface phosphate mine. Ideas to reduce risk factors were developed based on brainstorming sessions with those who had a stake in the task, i.e., those who performed the task, supervisors, safety representatives, and engineering and maintenance staff. NIOSH personnel facilitated the brainstorming sessions. It should be noted that all ideas generated during brainstorming are listed. The ideas for improvement are not listed in any particular order with regard to likelihood of successfully reducing risk factor exposure or ease of implementation. These ideas are unique to this mine site and the tasks observed and should not be generalized to all mine sites.

Table H–5.—Surface phosphate: float crew – wrenching pipes

Target task	Risk factors
Wrenching task (N=26) The float crew is responsible for most aspects of pipe maintenance and pipe relocation from one area of the mine to another.	Primary risk factors: • Handling of heavy and awkward materials • Repetitive hand and wrist movements • Hand and arm vibration from air tool usage • Awkward postures during pipe assembly tasks • Pinch-point exposures
\multicolumn{2}{c|}{Part of the body affected:}	
\multicolumn{2}{|l|}{Discomfort surveys were used. Workers cited low back, wrists, and shoulders. Pain was mild to moderate. Frequency of pain was monthly.}	
\multicolumn{2}{c|}{Ideas for improvement}	

- Standardize trucks, preferably a large size, which will provide more comfort for the crew while traveling the site and will allow for a more consistent set of tools and supplies to be available.
- Standardize equipment (wrenches, bars, air guns, etc.), and keep these on each float crew truck.
- Make pneumatic wrenches available to all float crews.
- Supply wrenchers with personal protective equipment such as impact gloves, knee pads, and extra-large rain suits that will allow them to perform their duties more efficiently.
- The air hose reel on all trucks should be mounted lower or between elbow and chest height.
- All bolt hole aligning tools should be equipped with a donut to prevent injury to hands and fingers when aligning pipes.

Table H–6.—Surface phosphate: hydraulic pit station operator – operating hydraulic pit station

Target task	Risk factors
Operating hydraulic pit station (N=3) Pit station operators are responsible for operating water cannons that break down mined material and filter debris.	Primary risk factors: • Repetitive joystick control usage • Excessive neck extension when viewing pump monitors • Contact stress – forearm and wrist • Awkward seated postures

Part of the body affected:

Discomfort surveys were used. Workers reported shoulder pain, mild to moderate, weekly.

Ideas for improvement

- Fully adjustable chairs would be better in the operator's compartment if the control panel would be redesigned to allow more adjustability.
- Compartments should be standardized so that operators can move from one location to another and work with the same setup.
- A new control layout design and new joysticks should be considered.
- Monitors need to be relocated to reduce neck stress. They should be adjustable to accommodate the wide range of users.

Table H–7.—Surface phosphate: dozer operation – moving material

Target task	Risk factors
Operating dozer (N=4) Dozers are used for pit preparation, assisting in pit setup, and prep work for pipelines.	Primary risk factors: • Repetitive control usage • Variable amounts of force required to operate controls • Whole-body vibration and jolting/jarring • Frequent twisting and turning of head and neck • Awkward seated postures
Part of the body affected: Discomfort surveys were used. All four workers reported neck and low back pain in which the pain severity was on average mild, but occurred weekly.	
Ideas for improvement	
• Proper standardized control handles should be used to reduce hand, arm, and shoulder stress. • Better mirror placement would eliminate some neck turning. • Additional training of mine workers to alert them of hazards of working around dozers and one-on-one training of dozer operators can minimize the hazards of working around dozers and relieve some of the stress experienced by dozer operators. • Replace or redesign the dozers to meet the following design issues: suspensions that result in less jolting, jarring, and bouncing, and equipped with better cutting blades.	

Table H–8.—Surface phosphate: reagent operator – emptying reagent cars

Target task	Risk factors
Emptying reagent cars (N=2) Reagent operators unload soda ash from carrier cars onto a belt, which takes it to the processing plant. Their job involves using heavy tools and working overhead.	Primary risk factors: • Handling heavy tools • Forceful, jerk-type exertions • Working with awkward upper-body postures • Working with hands above head • Uneven and slippery walking surfaces
Part of the body affected: Discomfort surveys were used. Wrist, hands, shoulders, neck, back, and knees were cited. Mild to moderate pain daily.	
Ideas for improvement	
• Provide operators with a hydraulic tool to turn the gate openings. • A vibrating mechanism inside the car or attached to the outside of the car would help to collect materials that stick to the inside of the cars. This mechanism would eliminate the need for pounding the cars with the hammer and rubber mallet. • A car puller cable system from one end of the car to the other could be developed that would move the cars from one bin door to the next. • Operators should make sure that their area is clear of debris and watch their footing. • Operators need to be more aware of their postures when performing this job. Aligning tools should be equipped with a donut to prevent injury to hands and fingers when aligning pipes.	

APPENDIX I.—UNDERGROUND LIMESTONE MINE

This mine is an underground limestone operation in the Eastern United States with 43 employees. The site consists of an underground room-and-pillar stone quarry and a stone processing plant. At the quarry, faces and backs are scaled, faces are drilled and blasted, shot stone is loaded into haul trucks, and stone is dumped into plant hoppers or raw material stockpiles. At the plant, stone may be crushed and is screened to acquire a final product. The final product is inspected and, if correct, loaded into haul trucks and added to the proper stockpiles.

The evaluation plan consisted of three steps. The first step was to select jobs for further study. A decision matrix was used to identify these jobs/work tasks. The matrix considered mine incident data, Nordic Questionnaire responses, supervisor interviews, and management concerns. Five work groups had a decision matrix score of 7 or higher out of a possible score of 10. They were the blasting crew (9), mechanics (9), scaling machine operators (8), quality control technicians (7), and plant laborers (7). It was determined that tasks performed by the laborer and mechanics vary too much for effective analysis. The blasting crew, scaling machine operators, and the quality control technicians were chosen for further study.

The second step was to select target work tasks within the blasting crew, mechanical scaling operations, and quality control technician jobs. Following work task observations and analysis of body part discomfort interviews, primary tasks were selected for each work group. They were hand scaling by the blasting crew, handling ANFO and dynamite by the blasting crew, scaling faces and backs by the scaling machine operators, and gathering samples by the quality control technicians.

The final step was to use video task analysis and brainstorming sessions with mine workers to characterize ergonomic risk factors and develop general risk reduction ideas. A brief overview of the baseline data, a copy of the decision matrix, risk factors identified, and general ideas for task improvement are presented below.

INCIDENT DATA

The mine provided nearly 3½ years of incident data from January 1, 1997, to May 31, 2000. Seventeen records were obtained. These incidents involved near-misses, medical treatment, and lost work time. Incident data were reviewed and summarized by the evaluation team. An incident ratio was calculated for each work group. Table I–1 shows the number of employees and an incident ratio for each work group. The incident ratio gives a relative estimate of the incident risk for each work group. The three groups with the highest incident ratio are the blasting crew, crusher operators, and mechanics. Table I–2 summarizes the 17 incidents by work group and incident type. Each incident narrative was reviewed and classified as one of five types: (1) jarred/bounced, (2) slip/fall, (3) caught between/struck by, (4) overexertion, or (5) other. The most common incident type, which accounted for more than half of the incidents (10 of 17, or 59%), was "caught between/struck by."

Table I–1.—Underground limestone: number of incidents and employees by work area, job title, and incident rate, January 1997–May 2000

Work area and job title	Incidents	Employees	Average annual incident rate[1]
Mine and plant:			
Supervisors	0	3	0
Haul truck operators	3	10	0.09
Mine:			
Driller operators	1	5	0.06
Blasting crew	3	4	0.21
High-lift/loader operators	1	4	0.07
Mechanics	3	4	0.21
Scaling machine operators	1	3	0.10
Water truck operator	0	1	0
Plant:			
Quality control technician	0	1	0
Laborer	0	1	0
Welder	0	1	0
High-lift/loader operators	0	3	0
Crusher operators	2	3	0.19
Total	17	43	0.11

[1]Average annual incident rate = reported incidents divided by estimated number of employees divided by 3 years.

NORDIC QUESTIONNAIRE

Questionnaire responses were evaluated by counting the number of body parts reported and the number of reports of discomfort in similar body parts for each work group. Table I–3 shows the number of body parts with discomfort reported by each work group. Forty of forty-three workers responded to the questionnaire. The site-wide average of reported body parts with discomfort was about three per person. The average number of body parts with discomfort reported by mechanics, the quality control technician, the water truck operator, and supervisors was somewhat higher. Table I–4 shows reported body part discomfort that was rated as "above average" or "below average" by work group. To be considered "above average," response percentages had to be at least 50% higher than the overall percentage for reported discomfort over the past year or the past week. Similarly, "below average" response percentages had to be at least 50% lower than the overall percentages for reported discomfort over the past year or the past week. If a body part does not appear in the table, it was considered to have an "average" rating of reported discomfort. Work groups that reported higher than average body part discomfort were mechanics, scaling machine operators, the water truck driver, the quality control technician, and supervisors.

SUPERVISOR INTERVIEWS

Eight work groups were identified as having physically demanding tasks by three site supervisors. Comparative ratings were assigned by the evaluation team and were based on consensus discussions. A rating can be "low," "medium," or "high." The blasting crew, mechanics, and the plant laborer were rated as having "high" physical demands. Table I–5 summarizes supervisor interview comments for these three work groups.

Table I-2.—Underground limestone: incidents by work group and incident type, January 1997–May 2000

Work group	Jarred/ bounced	Slip/fall	Caught between/ struck by	Over-exertion	Other	Total
Supervisors						0
Haul truck operators	2		1			3
Driller operators	1					1
Blasting crew		1	2			3
Mine high-lift/loader operators			1			1
Mechanics			2		1	3
Scaling machine operators			1			1
Water truck operator						0
Quality control technician						0
Plant laborer						0
Welder						0
Plant high-lift/loader operators						0
Crusher operators			1	1		2
Unknown	1		2			3
Total	4	1	10	1	1	17

Table I-3.—Underground limestone: number of reported body parts with discomfort by work group

Work group	No. of respondents	Mine	Plant	Supervisor	Total	Average
Supervisors	3			15	15	5.0
Haul truck operators	9	20	2		22	2.4
Driller operators	5	8			8	1.6
Blasting crew	4	8			8	2.0
Mine high-lift/loader operators	3	8			6	2.0
Mechanics	4	20			20	5.0
Scaling machine operators	2	9			7	3.5
Water truck operator	1	12			12	12
Quality control technician	1		11		11	11
Plant laborer	1		2		2	2.0
Welder	1		2		2	2.0
Plant high-lift/loader operators	3		1		1	0.3
Crusher operators	3		7		7	2.3
Total	40	85	25	15	125	3.1

Table I-4.—Underground limestone: Nordic Questionnaire responses by work group

Work group	Above average ratings	Below average ratings
Supervisors	Neck, wrists/hands, upper back, knees	E bows, hips/thighs.
Haul truck operators	None	Neck.
Driller operators	None	E bows, wrists/hands, hips/thighs, ankles/feet.
Blasting crew	E bows	Neck, upper back, hips/thighs, ankles/feet.
Mine high-lift/loader operators	E bows	Wrists/hands, hips/thighs, knees, ankles/feet.
Mechanics	Neck, shoulders, hips/thighs, knees	Ankles/feet.
Scaling machine operators	Neck, wrists/hands, upper back, knees	Shoulders, elbows, hips/thighs.
Water truck operator	All body parts	None.
Quality control technician	All body parts	None.
Plant laborer	Lower back and wrists/hands	Neck, shoulders, e bows, upper back, hips/thighs, knees, ankles/feet.
Welder	Lower back and knees	Neck, shoulders, e bows, wrists/hands, upper back, hips/thighs, ankles/feet.
Plant high-lift/loader operators	None	All body parts except lower back.
Crusher operators	Lower back	Shoulders, elbows, wrists/hands, hips/thighs.

DECISION MATRIX

Results of the decision matrix are shown in table I-6. Five work groups received a score of 7 or higher: the blasting crew (9), mechanics (9), scaling machine operators (8), quality control technician (7), and plant laborer (7). It was determined that tasks performed by the laborer and mechanics vary a lot from day to day. This meant that selecting and studying specific tasks would be difficult. For this reason, the blasting crew, scaling machine operators, and the quality control technician were chosen as the target work groups.

BODY PART DISCOMFORT INTERVIEWS

Four blasting crew members, three scaling machine operators, and one quality control technician were interviewed. The quality control technician had felt pain in his back and knees every day. Pain in his upper and midback had been mild, but the low back and knee pain had been moderate. He noted that recently he had felt unbearable pain in his upper legs and continued to have pain in this area. All members of the blasting crew interviewed had felt pain in their shoulders. Two members had felt pain in their elbows. The average pain severity in these two areas was mild to moderate, but the pain occurred at least a couple of times per week. In addition, the member of the blasting crew who fills ground holes reported daily severe or unbearable pain in his knees and ankles. All scaling machine operators had felt pain in their neck. Two operators had felt pain in all areas of their back and in their knees. Neck pain is felt nearly every week but is considered mild. Back pain is felt at least a couple of times per week (upper back pain is felt every day).

TASK-SPECIFIC RESULTS

Tables I-7 to I-10 summarize information gathered for the specific tasks studied at the underground limestone mine. Ideas to reduce risk factors were developed based on brainstorming sessions with those who had a stake in the task, i.e., those who performed the task, supervisors, safety representatives, and engineering and maintenance staff. NIOSH personnel facilitated the brainstorming sessions. It should be noted that all ideas generated during brainstorming are listed. The ideas for improvement are not listed in any particular order with regard to likelihood of successfully reducing risk factor exposure or ease of implementation. These ideas are unique to this mine site and the tasks observed and should not be generalized to all mine sites.

Table I-5.—Underground limestone: physically demanding jobs identified by supervisors

Work groups with physically demanding tasks	Key comments	Rating
Blasting crew	Hand scaling is physically demanding. Workers use a scaling bar to pull loose rock from faces and backs. A high degree of pushing and pulling force must be applied to the pry bar. They lift and carry 50- to 55- b bags and boxes of dynamite and ANFO. They clean out bottom holes at a drilled face with a handheld hoe while working on a rocky bottom.	High.
Mechanics	Workers handle heavy crib blocks, 250 lb each. They must lift, carry, and position the cr bbing as safety blocks before working on equipment. They do a lot of bending, lifting, and crawling around inside machines. This results in awkward postures and work in confined spaces.	High.
Plant laborer	75% production, 25% maintenance. The most physically demanding tasks are normal production cleanup duties. The worker is at risk of trips and falls on catwa ks, where there are lots of uneven, muddy, wet, or icy surfaces. He repeatedly uses a grease gun. During maintenance he changes screens and bearings, cuts and welds, and does a lot of lifting and carrying.	High.

Table I-6.—Underground limestone: decision matrix for selecting tasks

Work group	No. of employees	Incident data	Nordic Questionnaire	Supervisor interviews	Management concern	Final score
Supervisors	3	Low	High	Low		5
Haul truck operators	10	Medium	Medium	Low		5
Driller operators	5	Medium	Low	Low		4
Blasting crew	4	High	Medium	High	Yes	9
Mine high-lift/loader operators	4	Medium	Medium	Medium		6
Mechanics	4	High	High	High		9
Scaling machine operators	3	Medium	High	Medium	Yes	8
Water truck operator	1	Low	High	Low		5
Quality control technician	1	Low	High	Medium	Yes	7
Plant laborer	1	Low	Medium	High	Yes	7
Welder	1	Low	Medium	Low		4
Plant high-lift/loader operators	3	Low	Low	Low		3
Crusher operators	3	High	Medium	Low		6

Table I-7.—Underground limestone: quality control technician – gathering and testing samples

Target task	Risk factors
Gathering samples (N=1) A sample is shoveled into a bucket at a stockpile and driven back to the lab. The sample is then carried down 13 steps to the lab.	Primary risk factors: • Forceful exertions with shoulders and back when shoveling • Concentrated pressure points to hand from a power grip on the bucket handle • Lifting and carrying heavy loads, 40- to 70-lb buckets • Asymmetric load handling when carrying one bucket • Ascending and descending steps while carrying loads • Ascending and descending steps with hand(s) occupied
Testing sample (N=1) A sample is split, weighed, and dried, shaken for 10 min, and weighed again. When testing is done, the sample is dumped back into the bucket.	Primary risk factors: • Twisting, turning, and bending while handling sample pans • Lifting loads above shoulders to dump • Pulling trays and lifting loads from below knuckle height • Dust in the lab when analyzing a sample
Part of the body affected:	
Discomfort surveys were used. One worker reported pain. Average to moderate pain: upper, mid, and lower back; knees; neck; shoulders; and upper legs.	
Ideas for improvement	
• Use a shovel/spade that will push more easily into the sampling pile. • Park the truck as close as possible to the sampling pile to reduce carrying distances and worker exposure to equipment traffic. • Attach a platform to the rear of the truck to place samples in instead of into the truck bed. • Use soft rubber padding around the bucket handle to reduce concentrated pressure to hand. • Use a bucket size that holds a maximum of 40 lb without overfilling. • When collecting a heavier sample, shovel into two buckets. When carrying two buckets, keep the weight balanced and a maximum of 35 lb per bucket. • If possible, place the lab at ground level. This would eliminate the need to go up and down stairs and would significantly reduce materials handling. • If a ground level lab is not possible, develop an electric winch or other conveyance system that would transport samples down to and up out of the lab. • A better lab layout would improve process flow and reduce handling and twisting. • When dumping or transferring samples, try to keep the pan at hip height and do not raise the elbow above shoulder. • Antifatigue mats could help reduce aches and pain to lower extremities and lower backs.	

Table I-8.—Underground limestone: blasting crew – handling ANFO and dynamite

Target task	Risk factors
Handling ANFO and dynamite (N=4) The crew unloads bags of ANFO from a storage trailer and boxes of dynamite from a secure storage shed into the back of a truck. ANFO is then loaded into the hopper of the blaster's truck.	Primary risk factors: • Bending over, twisting, and reaching to lift ANFO in 50-lb bags and dynamite in 55-lb boxes • Carrying awkward and heavy loads • Loading ANFO into the hopper requires twisting and lifting with the elbow above shoulder
Part of the body affected:	
Discomfort surveys were used. Four workers reported pain. Average to moderate pain: shoulders, mid and low back.	
Ideas for improvement	
• Increase worker awareness of proper lifting and carrying techniques. This should include a problem-solving component and would include ergonomic principles that relate to blasting crew tasks performed at the site. • Work with suppliers to provide smaller ANFO and dynamite packaging, no more than 40 lb. • Modify the layout of ANFO and dynamite storage to reduce bending, twisting, and lifting distances. Consider spring-loaded pallets for ANFO bags. • A better, long-term solution would be to eliminate manual loading and unloading of ANFO. Store bulk material in a silo so that the ANFO can be directly loaded into the truck hopper.	

Table I-9.—Underground limestone: blasting crew – hand scaling and clearing bottom holes

Target task	Risk factors
Hand scaling (N=4) Before loading blast holes, two crew members are raised in a basket to knock and pry loose stone from the back and face using a pry bar.	Primary risk factors: • Forceful exertions with arms, shoulders, and back; the 8-ft pry bar creates a high force moment arm • Impact forces to hands, e bows, and shoulders when hammering with pry bar • Extended forward reaches when cleaning face • Working with arms above shoulder height • Working below the feet to remove material • Neck flexion when working with high and low targets • Wrist flexion/extension when prying material • Environment can be noisy, dark, smoky, foggy, and humid
Clearing bottom holes (N=4) After hand scaling, the ground man clears loose rock from the bottom holes using a hand shovel.	Primary risk factors: • Working on rocky, uneven ground, causing twisting and turning of ankles and knees • Forceful exertions with hands, arms, and shoulders • Working in kneeling or bent-over postures
Part of the body affected:	
Discomfort surveys were used. All four workers reported pain in the shoulders. Mild to moderate pain in the e bows. Clearing bottom holes often results in pain to the knees and ankles.	
Ideas for improvement	
• Try different types of scaling bars to reduce effort required. Ideas for improving the scaling bar: use an s-shaped bar that would require less movement when prying; use as much f berglass as possible with a thin metal center; try to counterbalance the bar. • Look for ways to eliminate hand scaling using some type of fail-safe extendable canopy. • The scaling basket should be repositioned more frequently to reduce reaches and poor postures. • Try to level and clean the floor before filling ground holes. A small scoop could be used after the hand scaling is done. • Keep the ground holes at least 3 to 4 in off the floor. • The ground man should use appropriate kneepads.	

Table I-10.—Underground limestone: scaling machine operator – mechanical scaling of face and back

Target task	Risk factors
Scaling face and back (N=3) The operator works two hydraulic joystick controls to scrape the tooth of the machine along the face or back, looking for loose stone. Once loose stone is found, he pokes and scrapes with the tooth to knock the stone free.	Primary risk factors: • Constant bouncing and jarring while scaling • The feet bounce on the floor of the compartment and against control stands • The operator uses repetitive hand and wrist movements to operate controls • Pushing and pulling on controls require frequent arm and shoulder exertions • The neck is flexed back when scaling the back • When scaling the back from the bench, you must lean forward to look up; the neck is not well supported
Part of the body affected:	
Discomfort surveys were used. Three workers reported pain. Average to moderate pain: neck, upper and mid back, knees, and forearms.	
Ideas for improvement	
• The ability of the seat to absorb shock is key. The newer seat that absorbs shock the best should be used in both machines unless a better seat is found. • There may be a better seat design. Consider the air seats used in some haulage trucks. • Any seat will wear out quickly in this situation. Evaluating and replacing the shock absorption system or the entire seat in a timely manner is important. • Improve the shock-absorbing characteristics of the operator compartment. One example would be to use shock-absorbing floor material. A long-term approach would be to work with an original equipment manufacturer (OEM) to build a machine with an isolated or free-floating compartment. • The seat should have a soft, adjustable support for the neck and head. • Consider testing alternate joystick controls to find one that would reduce hand and arm activity. • Regular breaks to move about and loosen up should be allowed. Rotating work crews using a split-shift approach would be the best way to reduce daily exposures to whole-body vibration. • The best long-term approach is to work with an OEM to develop a remote-controlled scaling machine. For this to work, you must consider the type of operator feedback needed and the type of controller that will work best and provide adequate lighting.	

www.ingramcontent.com/pod-product-compliance
Lightning Source LLC
Chambersburg PA
CBHW081801170526
45167CB00008B/3282